FLORA OF THE GUIANAS

Edited by

A.R.A. GÖRTS-VAN RIJN
&
M.J. JANSEN-JACOBS

Series A: Phanerogams

Fascicle 21

123. VOCHYSIACEAE
(L. Marcano-Berti)

123a. EUPHRONIACEAE
(L. Marcano-Berti)

124. TRIGONIACEAE
(E. Lleras)

126. KRAMERIACEAE
(B.B. Simpson)

including
Wood and Timber
(B.J.H. ter Welle, P. Détienne & N. Espinoza de Pernía)

1998
Royal Botanic Gardens, Kew

Contents

ISBN 1 900347 59 8

123. VOCHYSIACEAE

by

LUIS MARCANO-BERTI[1]

Trees or shrubs, rarely subshrubs; indument of simple, malpighioid (2-branched), or stellate hairs. Leaves simple, opposite, in whorls, sometimes scattered; generally petiolate; stipules present, small or represented by glands; blades entire or subundulate; penniveined, main secondary veins 16-30 per side or very numerous (up to 30 per cm), submarginal vein present or absent. Inflorescences: panicles, panicles of cincinni rarely of cymes, sometimes axillary cincinni. Flowers hermaphroditic, perigynous or epigynous, zygomorphic; sepals 5, imbricate, very unequal (subequal in *Salvertia*) inserted on a very short tube or hypanthium, posterior one always spurred, generally convolute, spur exserted or hidden in bud; petal 1, open or convolute, or petals 3 or 5, free, imbricate, sometimes absent, rudimentary petals sometimes present; stamen 1, anther innate or dorsifixed, bithecate, glabrous or pubescent, opening longitudinally; staminodes generally present; ovary superior and 3-locular or inferior and 1-locular, ovules 1 to numerous, style 1, simple, stigma 1. Fruits 3-locular, loculicidal capsules or 1-locular, indehiscent and generally winged; seeds often winged.

Distribution: A pantropical family of seven genera and ca. 222 species. Only *Erismadelphus* is paleotropical, the others are neotropical, five occur in the Guianas with 34 species.

[1] Herbarium Division, Department of Plant Ecology and Evolutionary Biology, Heidelberglaan 2, 3584 CS Utrecht, The Netherlands.

- Acknowledgements
The author is grateful to Prof. Dr. F. Weberling and Dr. T. Stützel, Ulm, Germany, Dr. A. Charpin, Genève, Switzerland, Dr. S. Barrier, Paris, France, and to colleagues at: VEN, B, BM, CAY, G, K, M, MO, NY, P, PORT, U, US for the assistance and hospitality given.
Many thanks to Drs J.C. Lindeman, A.R.A. Görts-van Rijn, M.J. Jansen-Jacobs (U) and S.A. Mori (NY) for their critical remarks, to Mrs. Eddy Durán for typing the manuscript and to Graciela Hintz for the drawings.
These studies have been supported by: the French Government, through its Embassy in Venezuela; the Deutscher Akademischer Austauschdienst Federal Republic of Germany; Universidad de Los Andes, Mérida, Venezuela; CONICIT, Venezuela; and the Missouri Botanical Garden, St. Louis, Missouri, U.S.A.

LITERATURE

Marcano-Berti, L. 1969. Un nuevo Género de las Vochysiaceae. Pittieria 2: 3-27.
Marcano-Berti, L. 1973. Vochysiaceae. In L. Aristeguieta, Familias y géneros de los árboles de Venezuela. Instituto Botánico. Caracas.
Marcano-Berti, L. 1981. Dos Nuevas Especies de Vochysia. Pittieria 9: 29-32.
Marcano-Berti, L. 1989a. Vochysiaceae: Novedades y Correcciones. Pittieria 18: 5-14.
Marcano-Berti, L. 1989b. Euphroniaceae: una nueva familia. Pittieria 18: 15-17.
Normand, D. 1966. Les Kouali, Vochysiacées de Guyane et leurs bois. Revue Bois et Forêts des Tropiques 110: 3-11.
Normand, D. 1967. Les Kouali, Vochysiacées de Guyane et leurs bois (suite et fin). Revue Bois et Forêts des Tropiques 111: 5-17.
Paula, J.E. 1969. Estudios sobre Vochysiaceae IV. Contribuiçao para o Vochysia Poiret e Erisma Rudge da Amazonia. Bol. Mus. Paraense Emilio Goeldi sér. Bot. 31: 1-23.
Stafleu, F.A. 1948. A monograph of the Vochysiaceae I. Salvertia and Vochysia. Rec. Trav. Bot. Néerl. 41: 397-546.
Stafleu, F.A. 1951. Vochysiaceae. In A. Pulle, Flora of Suriname 3(2): 178-199.
Stafleu, F.A. 1953. A monograph of the Vochysiaceae III. Qualea. Acta Bot. Néerl. 2(2): 142-217.
Stafleu, F.A. 1954. A monograph of the Vochysiaceae IV. Erisma. Acta Bot. Néerl. 3(4): 459-480.
Steyermark, J.A. 1966. Botanical Novelties in the region of Sierra de Lema, Venezuela. Bol. Soc. Venez. Ci. Nat. 110: 411-452.
Warming, E. 1875. Vochysiaceae. In C.F.P. von Martius, Flora Brasiliensis 13(2): 17-116.

KEY TO THE GENERA

1 Branchlets or leaves with simple or malpighioid hairs; ovary 3-locular, superior (= flowers perigynous); fruit dehiscent, not winged; seeds winged · 2
Branchlets or leaves with stellate hairs; ovary 1-locular, inferior or subinferior (= flowers epigynous or nearly so); fruit indehiscent, generally winged; seeds not winged · *1. Erisma*

2 Sepals subequal; spurred sepal not convolute; petals 5 · · · · · · · *4. Salvertia*
Sepals unequal; spurred sepal convolute in bud, enveloping inner flower parts; petals 0-3 · 3

3 Spurred sepal generally twice or less longer than the others; anther dorsifixed · 2. *Qualea*
Spurred sepal 3-4 times longer than the others; anther innate · · · · · · · · · · 4

4 Conspicuous glands representing stipules at base of petiole; petal 1, convolute; ovules 7 in each locule · · · · · · · · · · · · · · · · · · · 3. *Ruizterania*
No glands at base of petiole; petals 0-3, open or imbricate; ovules 2 in each locule · 5. *Vochysia*

1. **ERISMA** Rudge, Hist. Pl. Guiane 1:7. 1805.
Type: Erisma floribundum Rudge

Debraea Roem. & Schult., Syst. Veg. 1: 4, 34. 1817.
Ditmaria Spreng., Anleit. ed. 2. 2(2): 704. 1818, non Lühnemann, 1809.

Trees; indument of stellate hairs. Leaves opposite or in whorls of 3-4; stipules absent or small; main secondary veins 6-18 per side, submarginal vein present or absent. Inflorescences panicles or panicles of cincinni. Flowers epigynous or subepigynous; spurred sepal deciduous, up to twice as long as the others, convolute in bud, enveloping inner flower parts, spur exserted in bud, 4 smaller sepals unequal, persistent and accrescent in fruit; petal 1, white, yellow or violaceous, convolute in bud, enveloping the inner flower parts, generally obcordate, unguiculate, apex bilobed to emarginate, rudimentary petals present; stamen 1, anther dorsifixed, connective not cucullate; staminodes 0-4; ovary inferior or nearly so, 1-locular with 2 ovules, stigma terminal or lateral-terminal. Fruits indehiscent, 1-locular, generally winged by the enlarged smaller calyx lobes; seed 1.

Distribution: Ca. 19 species in Colombia, Venezuela, the Guianas, Peru and Brazil, represented by three species in the Guianas.

KEY TO THE SPECIES

1 Bracts persistent, ca. 10 x 8 mm; bracteoles persistent, stalked, oblanceolate, ca. 4.5 x 1.8 mm; spur incurved; petal yellowish; anther 2-3 mm long · 1. *E. floribundum* var. *floribundum*
Bracts deciduous, up to 5 x 2.5 mm; bracteoles deciduous, not stalked, linear to subulate, up to 1.7 x 0.3 mm; spur straight to uncinate recurved, not incurved; petal mainly white or violet; anther 1-2 mm long · · · · · · · · · · 2

2 Leaf glabrous on both surfaces; spur straight, pendent; petal mainly white · 2. *E. nitidum*
Leaf sparsely to densely stellate-pubescent below; spur uncinate-recurved; petal violet · 3. *E. uncinatum*

1. **Erisma floribundum** Rudge

In the Guianas only:
var. **floribundum**, Pl. Guian. 1: 7, t.1. 1805. – *Debraea floribundum* (Rudge) Roem. & Schult., Syst. Veg. 1: 34. 1817. – *Ditmaria floribunda* (Rudge) Spreng., Syst. Veg. ed. 16. 1: 16. 1825. Type: French Guiana, Martin s.n. (holotype BM, isotypes BR, M).

Erisma parvifolium Gleason, Bull. Torrey Bot. Cl. 60: 362. 1933. Type: Brazil, Mato Grosso, Krukoff 1401 (holotype NY, isotypes G, MO, P, U, US).
Erisma pallidiflorum Ducke, Arch. Inst. Biol. Veg. 2: 54. 1935. – *Erisma parvifolium* Gleason var. *pallidiflorum* Ducke, Arch. Inst. Biol. Veg. 4: 43. 1938. Type: Brazil, Amazonas, Ducke RB 24102 (holotype RB, isotypes K, NY, P, U, US).

Tree up to 35 m tall. Younger branchlets densely golden pubescent with short and some long stellate hairs (ca. 0.2 mm resp 0.5 mm long), older branchlets glabrous, cortex persistent. Leaves opposite or in whorls of 3-4; stipules absent; petiole 6-12 mm long, stellate-pubescent; blade ovate to elliptic, 8.5-16 x 2.5-6 cm, apex shortly acuminate, base obtuse to subrounded, glabrous above except primary vein with some stellate hairs, stellate-pubescent below; primary vein impressed above, prominent below, main secondary veins 12-17 per side, angle with primary vein 60°-80°, minor secondary veins 1-2 between each pair of main ones, veinlets numerous, forming a network of large, medium and small areolae above of large areolae below, submarginal vein 3.0-3.5 mm from the margin. Panicle ca. 20 cm long; cincinni up to 3 cm long, 2-5-flowered; bracts persistent, concave, subrounded, ca. 10 x 8 mm, stellate-pubescent; pedicels 1.2-1.4 mm long, densely stellate-pubescent; bracteoles persistent, stalked, oblanceolate, ca. 4.5 x 1.8 mm, densely stellate-pubescent. Spurred sepal, including calyx tube, ca. 8-9 mm long at anthesis, densely stellate-pubescent, spur erect-ascending, incurved, smaller sepals unequal, 2.5-5.5 mm long, densely pubescent; petal yellowish, glabrous, emarginate to bilobed; stamen glabrous, filament ca. 5 mm long, anther ca. 2-3 mm long; staminodes 2-4, glabrous, acicular, 0.3 mm long; ovary with stellate hairs at apex, style ca. 5.5 mm long, glabrous, stigma subcapitate, 0.45 x 0.3 mm. Fruit winged, ca. 6 cm long.

Distribution: Venezuela, French Guiana and Brazil; 6 collections studied (FG: 3).

Specimens examined: French Guiana: without locality, Martin s.n.; La Fumée Mt., Boom & Mori 1974 (NY); Wayampi village, Upper Oyapock R., Prévost & Sabatier 2846 (B, CAY, NY, P, U).

Phenology: Flowering from November to January; fruiting in December.

Note: The other var. *tomentosum* (Ducke) Stafleu is restricted to Central Amazonas, Brazil.

2. **Erisma nitidum** DC., Prodr. 3: 30. 1828. – *Qualea lutea* Martin, nomen nudum in syn. ex DC., Prodr. 3: 30. 1828. Type: French Guiana, Mt. Roura, Martin s.n. (holotype FI, isotype P).

Tree 5-8 m tall. Older branchlets glabrous, cortex persistent, cracked. Leaves opposite; stipules 2-3 mm long; petiole 8-10 mm long, glabrous; blade elliptic, 9-15 x 4.8-7.5 cm, apex obtuse to acuminate, base obtuse to truncate, glabrous on both surfaces; primary vein impressed above, prominent below, main secondary veins 7-9 per side, angle with primary vein 60°-70°, submarginal vein 5-7 mm from the margin. Inflorescence 24-30 cm long; cincinni up to 3 cm long, 3-6-flowered; bracts deciduous, ovate, ca. 3 mm long; pedicels ca. 1.0 mm long; bracteoles deciduous, linear, ca. 0.60 x 0.15 mm, pubescent. Spurred sepal, including calyx tube, 8 mm long at anthesis, densely minutely stellate-pubescent outside, inside covered with longer white simple hairs, spur 4.5-5.0 x 1.8-2.2 mm, straight, pendent, densely minutely stellate-pubescent, smaller sepals unequal, up to 5.6 mm long, stellate-pubescent; petal white with a yellow spot near the center, emarginate, pubescent above only on the lobules; stamen glabrous, filament ca. 5.5 mm long, anther ca. 2.1 mm long, apex acute and with mucro 0.1 mm long, base cordate; basal part of style pubescent. Fruit winged, up to 6 cm long.

Distribution: Guyana and French Guiana; 8 collections studied (GU: 3; FG: 5).

Selected specimens: Guyana: Potaro R., FD 3743 (K, U), Jenman 7439 (K, U), Abraham 347 (K, U). French Guiana: Oyapock R., Armontabo Cr., Oldeman B-1481 (P, U); Approuague R., Counamary Cr., Oldeman B-2834 (P).

Phenology: Flowering from October to November; fruiting from November to March.

3. **Erisma uncinatum** Warm. in Mart., Fl. Bras. 13(2): 110. 1875. Type: Brazil, Amazonas, Poeppig 2633 (lectotype W, isolectotypes BM, F, G, P, US). – Plate 1.

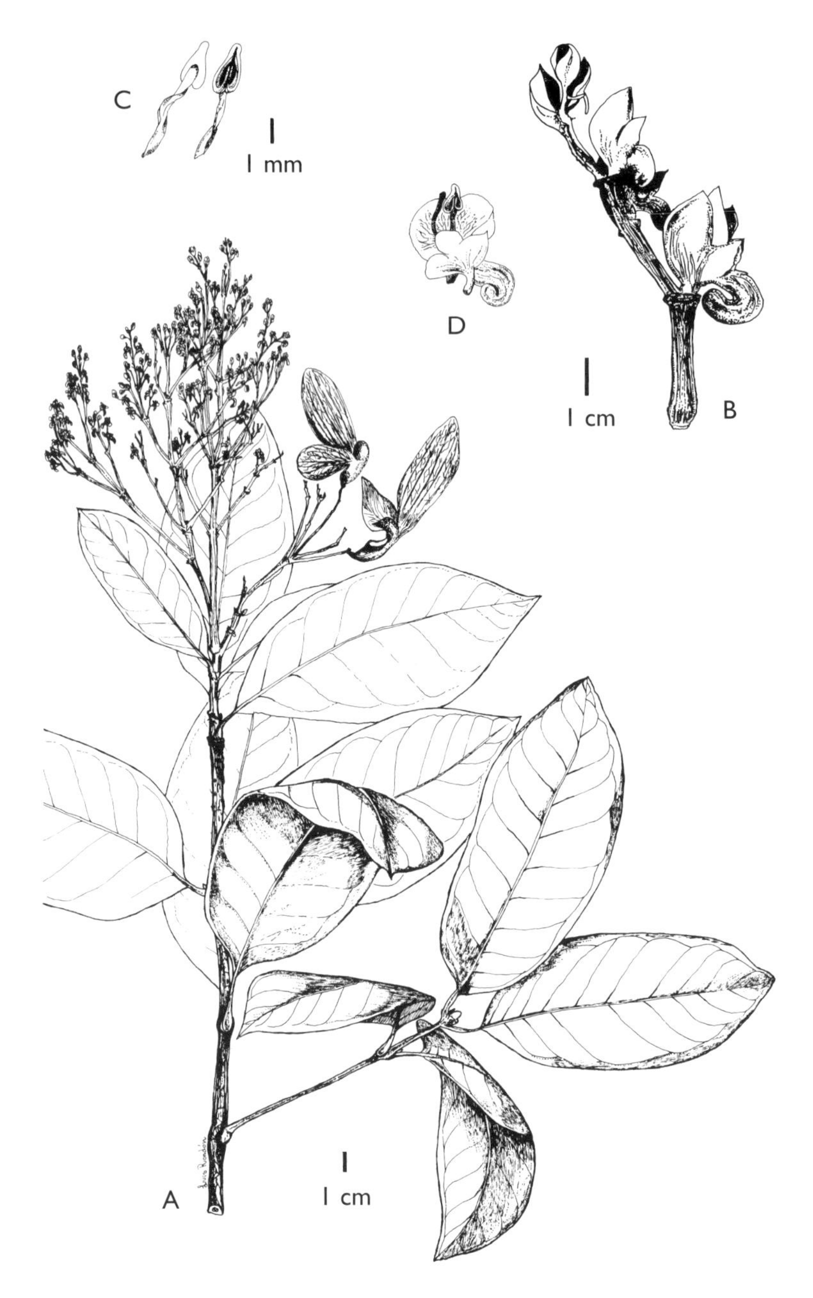

Plate 1. *Erisma uncinatum* Warm. A, flowering and fruiting branch; B, cincinnus; C, stamens, one seen on dorsal, other on ventral side; D, flower without petal. (Drawing by Luis Rondon).

Tree up to 40 m tall; stem up to 1.6 m diam. Branchlets stellate-pubescent, younger ones densely grey-pubescent, cortex persistent, cracked on older branches. Leaves opposite or in whorls of 3-4; stipules subulate-triangular, stellate-pubescent, ca. 2 x 0.6 mm, deciduous; petiole 10-20 mm long, stellate-pubescent; blade obovate to elliptic, 6.5-18 x 3-8.5 cm, margin slightly revolute, apex retuse-obtuse to retuse-truncate, base cuneate, glabrous above, sparsely to densely stellate-pubescent below; primary vein impressed above, prominent below, main secondary veins 6-9 per side at 1.5-2.7 mm distance, angle with primary vein 50°-60°, impressed above, subprominent below, veinlets numerous forming a network of large and small areolae, slightly evident above, rather conspicuous below. Panicle up to 40 cm long; cincinni up to 3 cm long, 3-5-flowered; bracts deciduous, ovate, 2.6-3.2 x 2.0-2.5 mm, acute to subobtuse at apex, truncate at base; pedicels 0.4-0.7 mm long, pubescent; bracteoles deciduous, subulate, 1.3-1.7 x 0.2-0.3 mm. Spurred sepal, including calyx tube ca. 7 mm long at anthesis, densely pubescent outside, 2-lobed: one lobe flat, pilose near the centre with hairs 0.4 mm long, other lobe subconcave, pubescent near border with hairs 0.1 mm long, spur uncinate-recurved ca. 4 x 4 mm, deciduous together with spurred sepal, smaller sepals ca. 4.5 mm long; petal violet, obcordate, emarginate, ca. 10 x 10 mm; stamen glabrous, filament complanate-subulate, 2.5-2.8 mm long and ca. 0.4 mm wide at base, anther 1.2-1.4 x 0.8 mm; connective produced beyond apex of thecae into a minute mucro; staminodes 4, aciculate, glabrous, 1 x 1 mm; style glabrous, stigma lateral-terminal, 0.5 mm long. Fruit 6-7.5 cm long, winged.

Distribution: Venezuela, Colombia, the Guianas, Peru and Brazil; 43 collections studied (GU: 2; SU: 8; SU: 12).

Selected specimens: Guyana: Kanuku Mts., FD 5801 (K), 5929 (K), Wilson-Browne 387 (K, U). Suriname: Natuurpark Brownsberg, LBB 13726 (U); Zanderij I, BW 3380 (U), 4751 (U). French Guiana: St. Laurent du Maroni, BAFOG 314M (P), 318M (P); Camopi R., Grenand 1362 (P).

Phenology: Flowering from May to December; fruiting from December to February.

Vernacular names: Guyana: prumaye. Suriname: iteballi beleru (Arowak); manoti-kwali (Paramacca); singrikwari (sranan-tongo); warapa-kwari. French Guiana: mam-onti-kouali, grignon fou.

2. **QUALEA** Aubl., Hist. Pl. Guiane 1: 5. 1775.
Type: Qualea rosea Aubl.

Amphilochia Mart., Nov. Gen. Sp. 1: 127. 1826.
Type: Amphilochia dichotoma Mart.
Agardhia Spreng., Syst. Veg. ed. 16. 1: 4. 1824.
Type: Agardhia cryptantha Spreng.
Schuechia Endl., Gen. Plant. 1178. 1840.
Type: Schuechia brasiliensis Endl. ex Walp.

Trees or shrubs; indument of simple hairs. Leaves generally oppposite, sometimes in whorls or rarely scattered; stipules generally represented by and/or provided with concave or subplane, elliptic to rounded, subprominent to cylindrical or urceolate prominent glands, sometimes normally developed, triangular, or present as stipular ridge, often provided with axillary glands; main secondary veins ca. 8 per side or very numerous (up to 30 per cm); submarginal vein present. Inflorescences generally panicles of cincinni, or rarely of regular cymes. Flowers perigynous; spurred sepal generally up to twice as long as the others, convolute and enveloping inner parts, spur hidden in bud, exserted or hidden at anthesis, 4 smaller sepals unequal; petal 1, white, yellow, pink, or blue, deciduous, often obcordate, convolute, enveloping inner flower part, rudimentary petals often present; stamen 1, anther dorsifixed, connective not or slightly produced; staminodes often present; ovary superior, 3-locular, with axile placentation, ovules up to 24 per locule, inserted in 2 rows, stigma terminal, subcapitate. Capsules 3-locular; seeds unilaterally winged.

Distribution: Ca. 48 species from Panama to Paraguay, 11 of which occur in the Guianas.

KEY TO THE SPECIES*

1 Petal mainly pink to blue 2
Petal mainly white to yellow 6

2 Spurred sepal, including calyx tube, 1.6-2.0 cm long at anthesis 3
Spurred sepal, including calyx tube, 0.5-0.9 cm long at anthesis 4

3 Leaves 6.5 x 3.6 cm, apex obtuse, base acute; stipular ridge absent; spur exserted at anthesis; petal dark blue . . . . *5. Q. mori-boomii*
Leaves 8-13 x 3.5-5 cm, apex acuminate, base obtuse to rounded; stipular ridge generally straight; spur hidden at anthesis by the spurred sepal or deciduous with it; petal pink . . . . *7. Q. polychroma*

* Note added in proof: Jansen-Jacobs et al. 5407 (K, U), a recent collection from Wakadanawa Savanna, Guyana, represents an as yet undescribed species of *Qualea*.

4 Main secondary veins ca. 14 per cm, with 3-4 minor ones inbetween ···· ··· *2. Q. coerulea*
Main secondary veins no more than 7 per cm and up to 16 per side with 1-2 minor ones inbetween ······································ 5

5 Main secondary veins 5-7 per cm; vegetative buds 1.8-2.5 mm long; spurred sepal, including calyx tube, up to 6 mm long at anthesis ··· *3. Q. dinizii*
Main secondary veins 7-10 per side; vegetative buds 3-4 mm long; spurred sepal, including calyx tube, 8.0-9.8 mm long at anthesis ············ ··· *8. Q. psidiifolia*

6 Main secondary veins 1-3 per cm; spur at anthesis 24-30 mm long ······ ·· *4. Q. grandiflora*
Main secondary veins more than 8 per cm; spur at anthesis less than 12 mm long ··· 7

7 Margin of leaf blade blackish when dry ··················· *9. Q. rosea*
Margin of leaf blade concolourous with blade ····················· 8

8 Main secondary veins 15-20 per cm; base of leaf blade truncate, rounded, or cordate ··································· *1. Q. acuminata*
Main secondary veins 9-13 per cm; leaf base acute, cuneate or obtuse ··· 9

9 Spurred sepal, including calyx tube, 21-28 mm long; connective produced beyond the thecae into an acute apiculum ····· *10. Q. schomburgkiana*
Spurred sepal, including calyx tube, 15-20 mm long; connective produced beyond the thecae into an emarginate or truncate apiculum ········ 10

10 Vegetative buds glabrous; petiole 6-10 mm long, glabrous; leaves 5-10 x 2.3-4 cm; connective produced beyond the thecae into an emarginate apiculum ···································· *6. Q. paraensis*
Vegetative buds densely pubescent; petiole pubescent, 5-6.5 mm long; leaves 5.5 x 2.5 cm; connective produced beyond the thecae into a truncate apiculum ······························ *11. Q. tricolor*

1. **Qualea acuminata** Spruce ex Warm. in Mart., Fl. Bras. 13(2): 40. 1875. Type: Brazil, Amazonas, near Panuré, Spruce 2612 (lectotype C, isolectotypes BM, BR, G, K, NY, P).

Qualea speciosa Huber, Bol. Mus. Paraense Hist. Nat. 3: 425. 1902. Type: Brazil, Pará, Aramá, Huber MG 1844 (holotype MG, isotype G).

Tree 3-20 m tall, stem 8-20 cm diam. Younger and older branchlets blackish, glabrous, cortex persistent. Leaves opposite; stipules 0.7-0.8 mm long, glabrous, deciduous, stipular glands in axil of stipules, slightly prominent, subrounded to erect-elliptic, 2.0-2.5 mm long, ca. 1.5 mm broad, stipular ridge absent; petiole 2-10 mm long, glabrous to sparsely pubescent; blade oblong to oblong-elliptic, 8-16 x 3-5 cm, margin flat, apex acuminate (acumen up to 1.5 cm long), base truncate, rounded to cordate, glabrous on both surfaces, except primary vein pubescent; primary vein pubescent on both surfaces, impressed above, prominent and winged below, main secondary veins 15-20 per cm, slightly prominent on both surfaces, minor secondary veins 1-2 between each pair of main ones, veinlets numerous towards margin forming a network of long and narrow areolae, submarginal vein at 0.3-0.6 mm from margin. Inflorescence an axillary or terminal panicle; cincinni 2-flowered; bracts similar to stipular glands; pedicels 5-10 mm long, densely pubescent. Spurred sepal, including calyx tube, 18-25 mm long at anthesis, appressed-pilose outside, spur exserted in anthesis, 6-10 mm long, smaller sepals unequal, 4-13 mm long; petal white with yellow and red spots, obcordate, 35-50 x 35-50 mm, emarginate, rudimentary petals 2.2-3.8 x ca. 0.3 mm or absent; stamen glabrous, anther 5.2-12 x 1.2-1.8 mm, apically recurved and acute; ovary globose, style pubescent on lower 1.0-4.8 mm. Fruit ca. 25 x 15 mm.

Distribution: Colombia, Venezuela, French Guiana, Peru, Brazil and Bolivia; 46 collections studied, only 2 from French Guiana: without locality, Leprieur 284, and s.n. (P).

Phenology: Flowering almost during the whole year; fruiting from February to May.

2. **Qualea coerulea** Aubl., Hist. Pl. Guiane 1: 7, 3: t. 2. 1775. Type: French Guiana, Aublet s.n. (holotype BM).

Tree up to 45 m tall, stem up to 80 cm diam. Younger branchlets quadrangular, reddish, glabrous, cortex persistent, older branchlets subterete, cortex deciduous in long plaques. Vegetative buds 1.6-1.9 mm long. Leaves opposite; stipules deltoid, ca. 1.0 x 1.1 mm, glabrous, glands, if present, in axil of stipules, subcylindrical, ca. 0.6 x 0.8 mm; petiole 4-7 mm long; blade elliptic, 6-9 x 3-4 cm, apex generally acuminate, base acute to obtuse, glabrous on both surfaces, except primary vein which is scarcely pubescent; primary vein impressed above, prominent and winged below, wings ca. 0.2 mm wide, ciliate, main secondary veins ca. 13 per cm, scarcely prominent on both surfaces, minor secondary veins 3-4 between each pair of main ones, veinlets numerous, forming a conspicuous network except near primary vein, submarginal veins 2: one at 0.4-0.5 mm from margin, as prominent as the main secondary veins, second one less prominent and much closer to margin. Inflorescence 16-18 cm long;

cincinni 2-4-flowered; pedicels 7-8 mm long, densely pubescent. Spurred sepal, including calyx tube, 7-9 mm long at anthesis, emarginate, densely pubescent outside, spur ca. 6 x 2 mm, exserted at anthesis, smaller sepals unequal, 4.5-6.0 mm long; petal mauve to blue with yellow at base, obcordate, 15-20 x 15-20 mm, emarginate, glabrous; staminal filament pubescent at 2 mm from the base with hairs of 0.1 mm long, anther 2.0-2.3 mm long, connective sparsely pubescent dorsally, as long as or scarcely longer than thecae, obtuse at apex; ovary globose, densely pubescent, style emerging abruptly from the ovary. Fruit ovoid, 5-6 x 2.8-3.4 cm.

Distribution: Suriname, French Guiana and Brazil; ca. 51 collections studied (SU: 13; FG: 26).

Selected specimens: Suriname: Macreabo, BBS 405 (U); Natuurpark Brownsberg, LBB 12578 (U). French Guiana: Camopi R., Oldeman 2595 (P); Approuague R., Poncy 181 (P).

Phenology: Flowering from August to January; fruiting from December to May.

Vernacular names: Suriname: gronfolo, laagland gronfolo; mawsikwari (Sranan); gonfolo-kwali, papakai-kwali (Paramacca). Guyana: ir(y)akopi, watra-kwari, woto-kwari (Carib); meniridan. French Guiana: bois marie, conaie, gonfolo rouge, grignon bonit.

3. **Qualea dinizii** Ducke, Arch. Jard. Bot. Rio de Janeiro 1: 49. t. 17, 19E. 1915. Type: Brazil, Pará, Ducke MG 7791 (lectotype MG, isolectotypes BM, G). – Plate 2 A-F.

Tree up to 40 m tall, trunk up to 1.5 m diam; cortex deciduous. Younger branchlets tetragonous, reddish, sparsely pubescent; cortex persistent, older branchlets subterete to subtetragonous with cortex deciduous. Vegetative buds 1.8-2.5 mm long, acute, pubescent. Leaves opposite to subopposite; stipules represented by subconcave glands, ca. 0.7 x 0.5 mm; petiole 2-4 mm long, sparsely pubescent; blade oblong to elliptic, 7-11 x 3.5-4.5 cm, apex acuminate, base obtuse to subrounded, never cordate, glabrous on both surfaces (adult leaves), except primary vein sparsely pubescent below; primary vein plane to subplane above, prominent below, main secondary veins 5-7 per cm, at 1.5-3.0 mm distance, plane above, slightly prominent below, main and minor secondary veins ca. 11 per cm, veinlets numerous forming a dense network with areolae of different shapes. Inflorescence an axillary panicle, 5-10 cm long; cincinni 2-flowered; peduncle ca. 0.5 mm long;

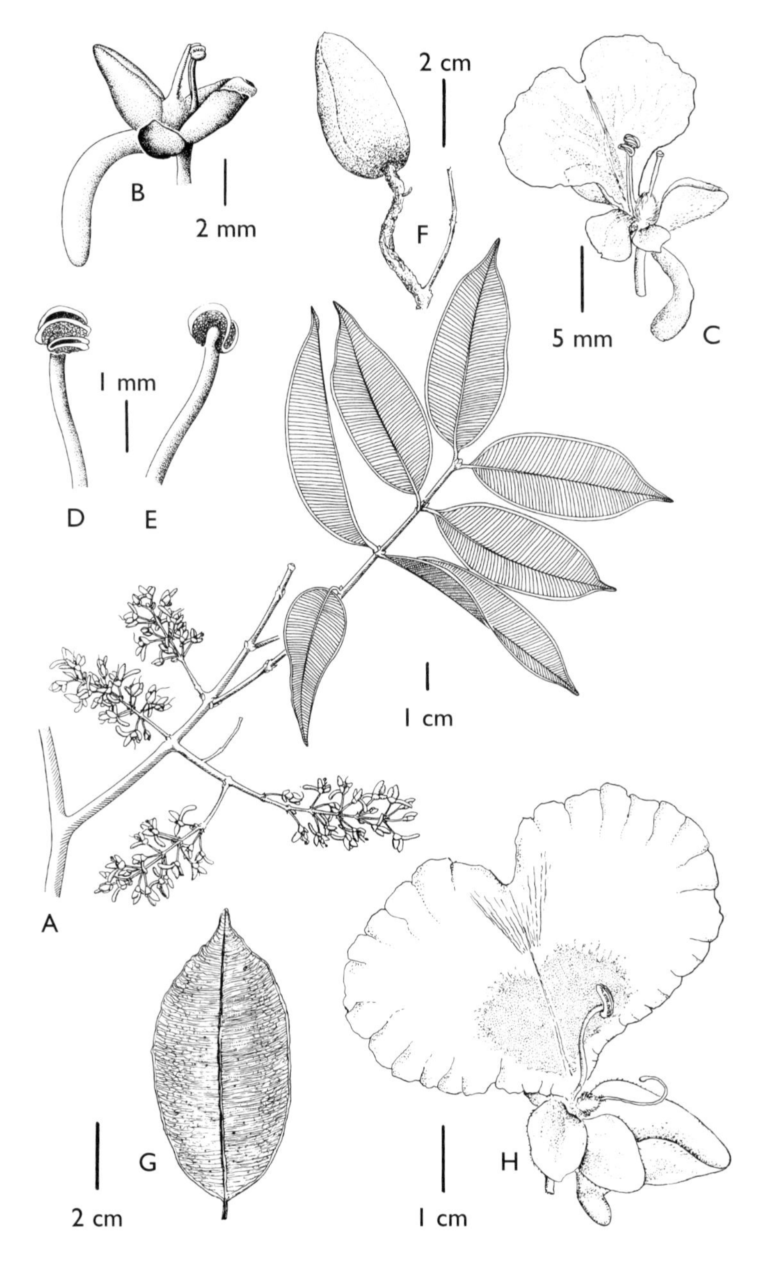

Plate 2. *Qualea dinizii* Ducke. A, flowering branch; B, flower without petal; C, open flower; D, stamen with open thecae; E, stamen showing insertion of filament to connective; F, fruit; *Qualea rosea* Aubl. G, leaf; H, open flower. (A, B, D, E by G. Hintz; C, G, H, by W.H.A. Hekking).

bracts similar to stipular glands; pedicels 3.7-6.0 mm long, sparsely pubescent. Spurred sepal, including calyx tube, 5.5-6.0 mm long at anthesis, densely greyish pubescent outside, spur exserted at anthesis, blackish, slightly to strongly recurved, greyish pubescent, smaller sepals unequal, 2-5 mm long, greyish pubescent outside; petal lilac to purple, obcordate, 100-130 x 13-18 mm, emarginate; stamen glabrous, filament subulate, 4.0-4.5 mm long x 0.5 mm wide at base and 0.2 mm wide at apex, anther recurved, 0.9-1.0 x 0.8-1.0 mm, connective not produced beyond apex of thecae; ovary densely pubescent, style glabrous. Fruit ovoid, 45-55 x 27-30 mm.

Distribution: Venezuela, the Guianas and Brazil; 26 collections studied (GU: 4; SU: 15 ; FG: 2).

Selected specimens: Guyana: Western extremity of Kanuku Mts., in drainage of Takutu R., A.C. Smith 3250 (B, NY); Powis Cr., Upper Courantyne R., FD 6771 (K). Suriname: Wilhemina Mts., Maguire et al. 53922 (NY, U); Wilhemina Mts., near Kayser airstrip, Irwin et al. 55974 (NY); Jodensavanne-Mapanekreek area, LBB 11197 (U). French Guiana: Saint Laurent du Maroni, BAFOG 6143 (P); Maroni R., Sabatier 991 (CAY, P).

Phenology: Flowering from December to October; fruiting from December to August.

Vernacular names: Guyana: yakopi (Carib); manau (Akawaio). Suriname: guyabakwari, waswaskwari (Sranan); guajavekwari (Suriname-Dutch); wosi-wosi (Carib).

4. **Qualea grandiflora** Mart., Nov. Gen. Sp. Pl. 1: 133. t. 79. 1824. Type: Brazil, São Paulo, Martius s.n. (holotype M).

Qualea ecalcarata Mart., Nov. Gen. Sp. Pl. 1: 131, t. 78. 1824. – *Schuechia brasiliensis* Endl. ex Walp., Rep. Bot. Syst. 2: 68. 1843. – *Schuechia ecalcarata* (Mart.) Warm., Vid. Meddel. Naturhist. Foren. Kjoebenhavn 33. 1867. Type: Brazil, Minas Gerais, Martius s.n. (holotype M).

Small tree or shrub, deciduous. Branchlets pubescent, cortex deciduous. Vegetative buds 3.0-5.0 mm long, obtuse-acute, densely pubescent. Leaves opposite; stipules represented by subprominent glands, 1.3-1.6 mm wide, below stipular glands accessory glands sometimes present, 0.6 mm wide; petiole 6-12 mm long, densely pubescent; blade discolorous, oblong or suboblong, 10-24 x 3.5-9 cm, apex acuminate to obtuse, base rounded, cordate to subcordate, glabrescent above, densely pubescent beneath; primary vein impressed above, prominent below, main secondary veins 1-3 per cm, impressed to subplane above, subprominent and wider below, minor secondary veins at least one between each pair of main secondary veins, veinlets numerous forming a network of rounded to quadrangular areolae, submarginal vein at 0.7-2.5 mm from margin. Inflorescence a terminal or axillary panicle or sometimes axillary cincinni; cincinni 1-5-flowered, up to 5.5 cm long, provided with glands similar to the stipular glands; peduncle 3.0-4.6 mm long, pubescent; pedicels 10-20 mm long, pubescent. Spurred sepal, including calyx tube, ca. 24-30 mm long at anthesis, densely pubescent outside, spur exserted at anthesis, slightly curved, 24-30 mm long, smaller sepals unequal, 13-21 mm long, densely pubescent outside; petal yellow, glabrous, obcordate-orbicular, ca. 40 x 60 mm; staminal filament 12-14 mm long, glabrous, anther 10-13 x 6.0-8.5 mm at base; staminodes 2, ca. 1 mm long; ovary spherical, pubescent, style emerging abruptly from the ovary, stigma lateral-terminal. Fruit ovoid, 8.0-12.0 cm long.

Distribution: Suriname, Peru, Brazil, Bolivia and Paraguay; ca. 100 collections studied, 1 from Suriname: Sipaliwini savanna area on Brazilian frontier, 4 km S of Meyers airstrip II, Oldenburger et al. 1158 (U).

Phenology: Flowering from October to March; fruiting from February to August.

5. **Qualea mori-boomii** Marc.-Berti, Pittieria 18: 5. 1989. Type: French Guiana, La Fumée Mt., Mori & Boom 15225 (holotype P, isotypes CAY, NY, U).

Tree 35 m high, stem 40 cm diam. Older branchlets glabrous, cortex persistent, cracked. Vegetative buds 1 mm long. Leaves opposite; stipules present, stipular glands subcylindrical, in axil of stipules, stipular ridge absent; petiole ca. 10 mm long; blade obovate, ca. 6.5 x 3.6 cm, margin subrevolute, apex obtuse, base acute, glabrous on both surfaces; primary vein impressed above, prominent and winged below, secondary veins 14-20 per cm, minor ones at least 1 between each pair of main ones; veinlets present below mainly towards margin, not evident above, submarginal vein at ca. 0.4 mm from margin, as thick and prominent as main secondary veins. Inflorescence ca. 10 cm long; cincinni subsessile, 2-flowered; pedicels ca. 7.2 mm long, sparsely pubescent. Spurred sepal dark blue, greyish appressed-pubescent outside, including calyx tube, ca. 20 mm long at anthesis, spur exserted at anthesis, appressed to the spurred sepal to subperpendicular to it, oblong to suboblong, 8.0-9.0 x 3.0-3.4 mm, smaller sepals unequal, 6-11 mm long; petal dark blue; staminal filament glabrous, anther pubescent dorsally, 5.2 mm long in bud; ovary pubescent, style emerging abruptly from ovary, stigma terminal.

Distribution: Known only from the type collection.

Phenology: Flowering in November.

Note: The type specimen, Mori & Boom 15225, was cited by Mori (1987) as *Q. coerulea*. In *Q. mori-boomii* spurred sepals are much larger, spurs longer and anthers larger than in *Q. coerulea*.

6. **Qualea paraensis** Ducke, Arch. Jard. Bot. Rio de Janeiro 1: 48, t. 16. 1915. Type: Brazil, Belém, Guédès MG 1591 (lectotype MG, isolectotypes BM, F, G, P, US).

Tree up to 30 m tall. Younger branchlets tetragonous, glabrous, reddish, older branchlets terete, cortex cracked or deciduous in plaques. Vegetative buds ca. 1.2 mm long, acuminate, glabrous. Leaves opposite; stipules 1 mm long, deltoid, stipular glands generally subcylindrical, 1.0-2.5 x 0.7-1.0 mm in axil of stipules, not always present, stipular ridge, if present, straight or arched; petiole 9-13 mm long, glabrous, subterete below; blade generally elliptic, 5.5-10 x 2.3-4 cm, apex acuminate (acumen 5-10 mm long), base obtuse, glabrous on both surfaces, except primary vein sometimes pubescent; primary vein impressed, glabrous to glabrescent above, prominent, narrowed, 2-winged below, wings 0.1 mm wide, sparsely ciliate, main secondary veins 9-12 per cm, slightly prominent on both surfaces, minor secondary veins 1-3 between each pair of main secondary veins, narrower and shorter than the former, veinlets numerous, forming a network of narrowed and long areolae perpendicular to primary vein, submarginal vein at ca. 0.5 mm from margin. Inflorescence a terminal or axillary, 3-5 cm long panicle; cincinni 2-flowered; bracts similar to stipules; pedicels 6-10 mm long, pubescent. Spurred sepal, including calyx tube, 15-18 mm long at anthesis, densely pubescent outside, spur 4.5-7.3 x 2.8-3.9 mm, exserted at anthesis, smaller sepals unequal 4.0-10.5 mm long, pubescent outside; petal white with yellow or red spots above, ca. 40 mm long and wide; staminal filament glabrous or sparsely pubescent, anther 7-9 mm long, connective pubescent dorsally, produced beyond apex of thecae into an emarginate apiculum 1.3-3.0 mm long; ovary globose, pubescent, style glabrous except basal 1.0-1.5 mm.

Distribution: Colombia, Venezuela, Guyana, Peru and Brazil; 10 collections studied, 2 from Guyana: Berbice-Demerara watershed, FD 832 (K); Kassikaityu R., Upper Essequibo R., Myers 5662 (K).

Phenology: Flowering from September to May.

7. **Qualea polychroma** Stafleu, Acta Bot. Néerl. 2(2): 182, fig. 11. 1953. Type: Guyana, Mt. Roraima, Arabupu, FD 2832 (holotype K).

Tree 25 m tall. Younger branchlets tetragonous, blackish, cortex persistent like on older branchlets. Vegetative buds 0.7 mm long, truncate. Leaves opposite; stipules deltoid, ca. 1.7 x 2.0 mm, apex acute, deciduous, stipular ridge generally straight; petiole 10-12 mm long, glabrous; blade elliptic to elliptic-oblong, 8-13 x 3.5-5 cm, apex acuminate, base obtuse to rounded, glabrous on both surfaces; primary vein impressed to subimpressed above, prominent, 2-winged below, main secondary veins 11-16 per cm, arched, slightly prominent, minor secondary veins 1-2 between each pair of main ones, veinlets numerous, forming a network of long and narrowed areolae perpendicular to primary vein, submarginal vein at ca. 0.5 mm from margin. Inflorescence 14-16 cm long, terminal; cincinni 2-flowered; peduncle ca. 1 mm long; bracts similar to stipules; pedicels 13-15 mm long, pubescent. Spurred sepal, including calyx tube, 16-18 mm long at anthesis, appressed-pubescent, revolute and concealing spur at anthesis, sometimes deciduous, spur ca. 7 x 3 mm, oblong, hidden by spurred sepal at anthesis or both deciduous, smaller sepals unequal, 9.5-14 mm long; petal pink, glabrous, 35-45 x 35-45 mm, emarginate, rudimentary petals glabrous, 1.6-4.2 x 0.7-20 mm; staminal filament ca. 13 mm long, densely pubescent, anther oblong, 5.7-6.2 x 1.4-2.0 mm, pubescent dorsally, connective produced beyond apex of thecae into an obtuse apiculum, 0.60-0.68 mm long; staminodes glabrous, ca. 1.2 x 0.4 mm; ovary globose, densely pubescent, style pubescent on the lower 1.6 mm.

Distribution: Venezuela and Guyana; 4 collections studied. 2 from Guyana: the type collection, and Upper Mazaruni R. region, Kamarang, trail W of airstrip, Boom et al. 8421 (NY).

Phenology: Flowering from December to April.

8. **Qualea psidiifolia** Spruce ex Warm. in Mart., Fl. Bras. 13(2): 46, t. 8, fig. 1. 1875. Type: Venezuela, Terr. Fed. Amazonas, Casiquiare, Vasiva and Pacimoni Rs., Spruce 3059 (lectotype C, isolectotypes G, K, NY, P, U).

Tree up to 40 m high. Younger branchlets subtetragonous, reddish, cortex persistent, older branchlets subterete, glabrous, cortex persistent. Vegetative buds 3-4 mm long, glabrous. Leaves opposite or subopposite; stipules represented by subconcave, subprominent, subrounded to subelliptic glands, 1.4 x 0.5-1 mm, below the stipular glands accessory glands often present; petiole 5-12 mm long, glabrous to sparsely pubescent; blade elliptic to subelliptic, 4.5-18.0 x 3.5-6.0 cm, apex acute to obtuse, base obtuse to rounded or cordate, sometimes oblique, glabrous on both surfaces, sometimes primary vein sparsely pubescent below; primary vein impressed above, prominent below, main and minor secondary veins 14-16 per side at 1.0-3.5 cm distance, main secondary veins 7-10 per side, subprominent below, flat to slightly prominent above, at angle of 55°-65° with primary vein, veinlets visible on both surfaces, numerous, forming a network of rounded to quadrangular areolae, submarginal vein present at least near apical half at 2.0-4.5 mm from margin. Inflorescence a terminal panicle of cymes, up to 10 cm long; peduncles of basal cymes up to 5 mm long, sparsely pubescent; pedicels 4.0-6.2 mm long, pubescent; spurred sepal, including calyx tube, ca. 9.5 mm long at anthesis, pubescent, spur subspatulate, ca. 7.5 mm long, subrounded at apex, exserted at anthesis, smaller sepals 3.2-7.0 mm long; petal pink, obcordate, ca. 25 x 25 mm, glabrous, emarginate, unguiculate; staminal filament glabrous, ca. 9 mm long, anther curved, 1.8-2.0 mm long, obtuse at apex, glabrous; staminodes and rudimentary petals absent; ovary densely pubescent, style glabrous.

Distribution: Colombia, Venezuela, French Guiana and Brazil; 6 collections studied, 2 from French Guiana: Trois Sauts, Grenand 1128 (CAY); Prévost & Sabatier 2755 (B, CAY, U).

Vernacular name: French Guiana: kwali-sili (Wayampi); kuluwa.

Phenology: Flowering from September to October.

9. **Qualea rosea** Aubl., Hist. Pl. Guiane 1: 5; 3: t. 5. 1775. Type: French Guiana, Aublet s.n. (holotype BM, not seen). – Plate 2 G, H.

Qualea melinonii Beckmann in Engl., Bot. Jahrb. Syst. 40: 280. 1908. Type: French Guiana, Mélinon (1863) s.n. (BM, P, US).

Tree 20-34 m high, trunk 38-50 cm diam. Younger branchlets blackish, glabrous, cortex persistent, older branchlets with cracking cortex or deciduous in tiny plaques. Leaves opposite; stipules deltoid, ca. 1 mm long, acute, with axillary subcylindrical glands 1.2-2.0 mm long, ca. 1 mm diam., stipular ridge, if present, straight; petiole 8.5-14 mm long, glabrous; blade generally elliptic, 6.5-11.0 x 3.5-6.0 cm, apex acuminate, acumen to 1.3 cm long, base obtuse or subrounded, blackish between submarginal vein and margin when dry, glabrous to glabrescent on both surfaces; primary vein impressed and sparsely pubescent above, subprominent and 2-winged below, wings ca. 0.2 mm wide, ciliate along margin, main secondary veins 11-14 per cm, minor secondary veins 2-4 inbetween each pair of main ones and ca. half as long, veinlets forming a network of wide areolae, submarginal vein at 0.6-0.7 mm from margin, marginal vein present. Inflorescence few-flowered, ca. 6 cm long; cincinni 1-2-flowered; bracts resembling stipules; pedicels 8-15 mm long, pubescent. Spurred sepal, including calyx tube, 16-20 mm long at anthesis, suborbicular, emarginate, appressed-pubescent outside, spur broadly elliptic, 4.0-7.5 x 3.0-5.0 mm, sparsely pubescent, subrounded at apex, hidden in bud, exserted at anthesis, smaller sepals 3.5-10.5 mm long; petal white with reddish pink at the base and yellow at the middle, glabrous, obcordate, ca. 60 x 60 mm, rudimentary petals and/or staminodes linear; staminal filament glabrous, ca. 10 mm long, anther 8.0-8.5 mm long, connective glabrous or with some hairs dorsally and produced beyond the thecae into an acute apiculum 2.5-3.0 mm long, subrecurved towards the back; ovary subspherical, densely pubescent, style pubescent only at the 1.5 mm of the base, emerging abruptly from the ovary. Capsule up to 90 mm long.

Distribution: E Guyana (acc. to Fanshawe), Suriname and French Guiana; 40 collections studied (SU: 7; FG: 33).

Selected specimens: Suriname: Mapane, LBB 11182 (U); Natuurpark Brownsberg, LBB 14760 (U). French Guiana: route de Cayenne, BAFOG 7739 (P); 7747 (P).

Vernacular names: Guyana: yakopi (Carib); Suriname: bergigronfolo (Sranan); gonfolo-kwali (Paramacca); iryakopiran (Carib); meniridan. French Guiana: cèdre gris, cèdre jaune, gonfolo kouali, gonfolo rose.

Phenology: Flowering from September to December; fruiting from December to April.

10. **Qualea schomburgkiana** Warm. in Mart., Fl. Bras. 13(2): 39. 1875. Type: Guyana, Roraima, Schomburgk 893 (lectotype GH).

Tree up to 20 m tall, stem 20-30 cm in diameter. Younger branchlets tetragonous, glabrous, cortex persistent, blackish to red, older branchlets subterete, robust, glabrous, cortex persistent, cracked. Vegetative buds

1.5 mm long. Leaves opposite; stipules subdeltoid, ca. 0.9 mm long, glabrous, stipular glands concave, not prominent, erect, elliptic, ca. 2.5 x 1.5 mm, stipular ridge, if present, straight, arched or V-shaped; petiole 6-10 mm long, glabrous; blade lanceolate, ovate to elliptic-oblong, 6.2-12 x 2.5-5 cm, margin slightly revolute, apex acute, obtuse to acuminate with tiny mucro, base acute to obtuse, glabrous on both surfaces; primary vein glabrous on both surfaces, plane to impressed above, prominent and 2-winged below, wings 0.32-0.80 mm wide, main secondary veins 10-13 per cm, slightly prominent on both surfaces, minor secondary veins 1-3 between each pair of main ones, veinlets numerous near primary vein, forming a network with long, narrow areolae perpendicular to primary vein, submarginal vein at ca. 0.5-0.8 mm from margin, marginal vein parallel. Inflorescence an axillary panicle, 7-9 cm long, few-flowered, glabrous to glabrescent; cincinni 2-flowered; peduncle 1 mm long; bracts similar to stipules; pedicels 10-15 mm long. Spurred sepal, including calyx tube, 21-28 mm long at anthesis, densely pubescent outside, spur exserted at anthesis, 7-11 x 3.3-3.6 mm, smaller sepals unequal, 5.6-16 mm long; petal white, spotted with dark red and yellow, obcordate, 40-50 mm long, rudimentary petals up to 5 mm long; stamen glabrous, anther 6.0-7.5 mm long, connective produced beyond the thecae into an acute apiculum; ovary densely pubescent, ovoid, style pubescent at basal 3.0-4.5 mm. Fruit 4.0-4.5 x 3.0-3.3 cm.

Distribution: Venezuela, Guyana and Brazil; ca. 30 specimens studied (GU: 9).

Selected specimens: Guyana: Membaru-Kurupung tail, Maguire & Fanshawe 32440 (NY); Haieka savanna, Upper Mazaruni R., Tillett et al. 45222 (NY).

Phenology: Flowering from August to April; fruiting from September to May.

Vernacular name: Guyana: manaw (Akawaio).

11. **Qualea tricolor** Benoist in Lecomte, Not. Syst. (Paris) 3: 176. 1915.
Type: French Guiana, Gourdonville, Benoist 1564 (holotype P).

Large tree. Younger branchlets reddish, cortex persistent, older branchlets with cortex deciduous in small plaques. Vegetative buds ca. 1 mm long, densely pubescent. Leaves opposite; stipules deltoid, ca. 0.8 mm long, glands, if present, subcylindrical, ca. 1.2 x 0.8 mm, stipular ridge absent; petiole 5-6.5 mm long, densely pubescent; blade elliptic, 5.0 x 2.5 cm, apex acuminate, base cuneate, glabrous on both surfaces, except primary vein; primary vein subimpressed and pubescent above, prominent, sparsely pubescent and narrowly winged below, main secondary veins 11-13 per cm, plane to slightly prominent on both surfaces, minor secondary veins 1-3 between each pair of main ones, forming with veinlets a network of long areolae, submarginal vein as prominent as main lateral veins, at 0.4 mm from margin. Inflorescence a terminal or axillary panicle; cincinni 1-2-flowered; pedicels sparsely pilose, 9-13 mm long. Spurred sepal, including calyx tube, 17-20 mm long at anthesis, densely pubescent outside, spur exserted at anthesis, oblong to elliptic, 7.0-10.5 x 2.7-2.9 mm, smaller sepals unequal, 7-13 mm long; petal white with yellow and red spots, obcordate-suborbicular, 40 x 40 mm, rudimentary petals 3.0-5.0 mm long; filament glabrous, anther 8.0-9.0 x 1.5 mm, connective pubescent dorsally and produced beyond apex of thecae into a truncate apiculum 1.1-1.8 mm long; ovary densely pubescent, style glabrous.

Distribution: French Guiana; 2 collections studied: the type collection; and without locality, Mélinon 142 (P).

Phenology: Flowering in August.

3. **RUIZTERANIA** Marc.-Berti, Pittieria 2: 6. 1969.
Qualea, series I *Calophylloideae* Warm. p.p., in Mart., Fl. Bras. 13(2): 30. 1875.
No type mentioned.
Qualea, sect. *Trichanthera* Stafleu, Acta Bot. Néerl. 2(2): 153. 1953.
Type: Ruizterania trichanthera (Spruce ex Warm.) Marc.-Berti (= Qualea trichanthera Spruce ex Warm.)

Trees or shrubs; indument of simple hairs. Leaves opposite, rarely subopposite; stipules represented by crateriform, rounded to elliptic, subprominent glands; primary vein narrowly winged below, secondary veins up to 50 per cm, submarginal vein present. Inflorescences axillary, 1-3-flowered cincinni and/or terminal or axillary panicles. Flowers perigynous; sepals inserted on a very short calyx tube or hypanthium, posterior one spurred dorsally at the base, 3-4 times longer than the others, convolute, enveloping inner flower parts, spur exserted in bud, smaller ones subequal; petal 1, white spotted with yellow to orange, convolute, enveloping inner flower parts, rudimentary petals sometimes present; stamen 1, anther innate, connective not cucullate, on one or both thecae hairs arranged like the bristles of a brush; staminodes often present; ovary superior, 3-locular, ovules 7 per locule, inserted in two rows, stigma terminal. Capsules 3-locular, 3-valved; seeds winged.

Distribution: Ca. 13 species in Venezuela, the Guianas, Colombia, Brazil and Peru; three species occur in the Guianas.

KEY TO THE SPECIES

1 Flower buds and pedicels glabrous or very sparsely greyish pubescent; ovary white to greyish pubescent · *1. R. albiflora*
Flower buds and pedicels densely ferruginous-pubescent; ovary ferruginous-pubescent · 2

2 Spurred sepal at anthesis 12-18(-20) mm long; leaf glabrous to densely ferruginous-pubescent below, base obtuse to subrounded, sometimes rounded; anther at anthesis (2.7-)5.0-9.0(-10.0) mm long · *2. R. ferruginea*
Spurred sepal at anthesis (20-)23-26 mm long; leaf glabrous to glabrescent below, base rounded to cordate at the same branch; anther at anthesis (9.0)10-13 mm long · *3. R. rigida*

1. **Ruizterania albiflora** (Warm.) Marc.-Berti, Pittieria 2: 9. 1969. – *Qualea albiflora* Warm. in Mart., Fl. Bras. 13(2): 36. 1875. Type: Suriname, Kappler 2037 (lectotype W, isolectotypes F, G, NY, L, P, U). – Plate 3 A, B.

Qualea glaberrima Ducke, Arch. Jard. Bot. Rio de Janeiro 1: 46. t. 19F. 1915. Type: Pará, Ducke MG 15491 (holotype MG, isotypes BM, F, G, P, US).

Tree to 35 m high, stem 80 cm diam. Younger branchlets sparsely grey-pubescent, cortex persistent, older branchlets glabrous, cortex cracked. Vegetative buds 1.0-1.5 mm long, grey-pubescent. Leaves opposite; stipular glands concave slightly prominent, elliptic to subrounded, 0.75-1.06 x 0.56 mm; petiole 3.5-10 mm long, glabrous to glabrescent; blade obovate to elliptic, 3-8.7 x 2-3.1 cm, margin flat to subrevolute, apex obtuse-retuse to subrounded-retuse with mucro 0.3-0.6 mm long, base acute to subrounded, glabrous above, generally glabrous below; primary vein glabrous, impressed to plane above, glabrous to glabrescent, prominent and 2-winged below, wings narrow, ciliate, main secondary veins 10-14 per cm, veinlets few, forming a network with long areolae near the margin, submarginal vein at 0.3-0.5 mm from margin. Inflorescence axillary, 2-4-flowered cincinni or/and panicle of cincinni; pedicels 4.5-11.0 mm long, glabrous to sparsely grey-pubescent. Flower bud grey-pubescent; spurred sepal, including calyx tube, 13-18 mm long at anthesis, margin ciliate, sparsely grey-pubescent outside, spur 7.5-9.0 x ca. 3.0 mm, generally clavate, smaller sepals 2.5-4.0 mm long; petal

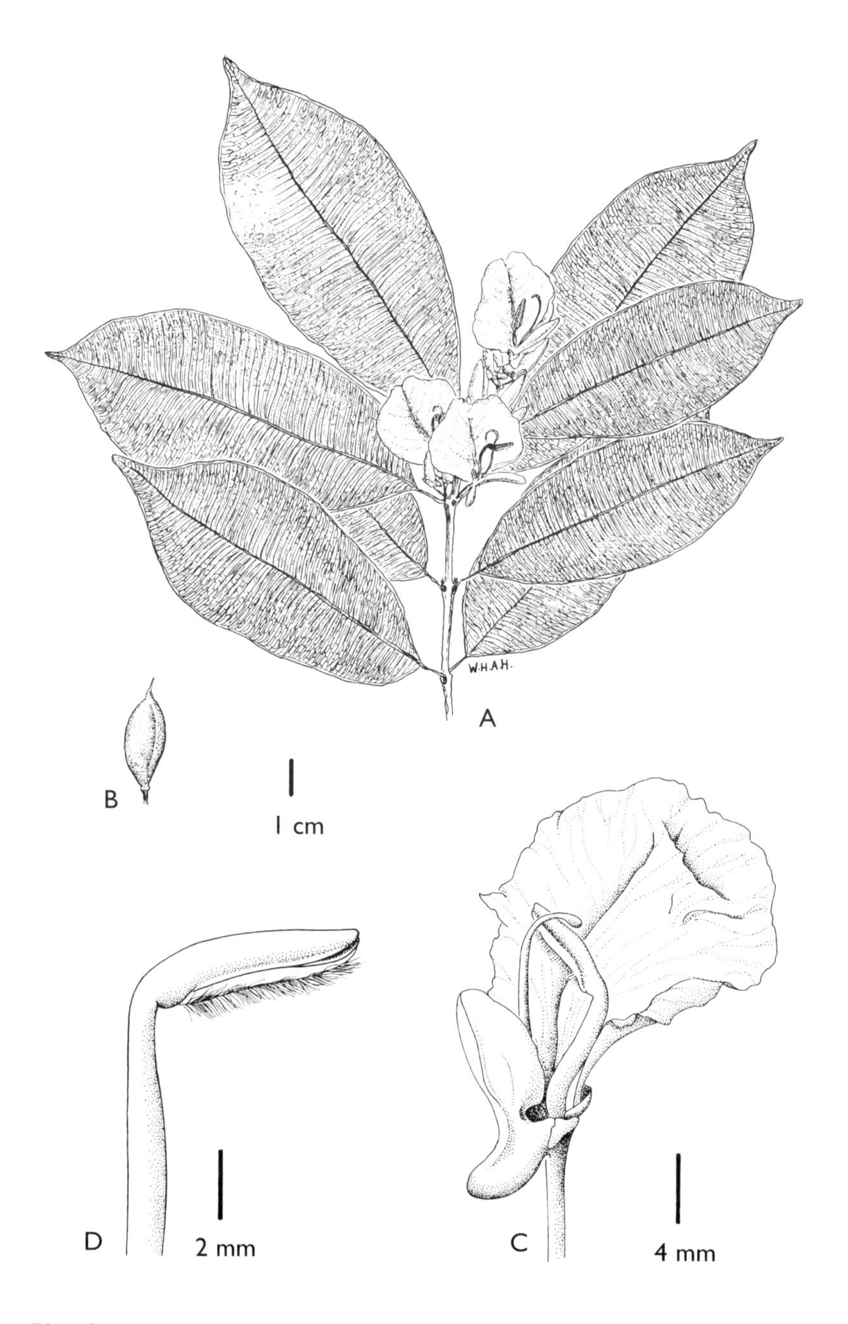

Plate 3. *Ruizterania albiflora* (Warm.) Marc.-Berti. A, flowering branch; B, fruit; *Ruizterania ferruginea* (Steyerm.) Marc.-Berti. C, flower; D, stamen. (A, B by W.H.A. Hekking; C, D by G. Hintz).

white spotted with yellow, 20-32 x 15-32 mm, rudimentary petals ca. 4 x 1 mm; staminal filament glabrous, 6.7-11.0 mm long, anther 6.5-11.0 mm long, one thecae glabrous to glabrescent with short and subappressed hairs, other theca densely pubescent with erect, long hairs (± 0.8 mm long); staminodes to 1 mm long; ovary grey-pubescent with hairs of ca. 0.3 mm long, style pubescent on the basal 1/5.

Distribution: The Guianas and Brazil; 17 collections studied (GU: 1; SU: 5; FG: 7).

Selected specimens: Guyana: Essequibo R., A.C. Smith 2709 (G, U). Suriname: Tafelberg, Maguire 24841 (NY); Jodensavanne, LBB 11180 (U). French Guiana: route de Cayenne, BAFOG 7572 (P); route de Mana, BAFOG 7722 (P).

Phenology: Flowering from June to December.

Vernacular names: Guyana: manau (Akawaio). Suriname: eigron-gronfolo (Sranan); hoogland gronfolo (Suriname-Dutch); ir(y)akopi, tupura iryakopi, yakopi (Carib); gonfolo-kwali (Paramacca); meniridan hohorodikoro. French Guiana: gronfolo-kouali.

2. **Ruizterania ferruginea** (Steyerm.) Marc.-Berti, Pittieria 2: 11. 1969. – *Qualea ferruginea* Steyerm., Fieldiana Bot. 28(2): 295. 1952. Type: Venezuela, Edo. Bolívar, Kavanayén Cr., between Santa Teresita de Kavanayén and airport, Steyermark 60914 (holotype F, isotypes A, NY, US, VEN). – Plate 3 C, D.

Qualea rubiginosa Stafleu, Acta Bot. Néerl. 2 (2): 154, fig. 4. 1953. – *Ruizterania rubiginosa* (Stafleu) Marc.-Berti var. *rubiginosa*, Pittieria 2: 14. (figs. 1,4,5). 1969. Type: Venezuela, Edo. Bolívar, Uaiparí R., Cardona 1905 (holotype US, isotype VEN).
Qualea rubiginosa Stafleu var. *angustior* Steyerm., Acta Bot. Venez. 2: 239. 1967. – *Ruizterania rubiginosa* (Stafleu) Marc.-Berti var. *angustior* (Steyerm.) Marc.-Berti, Pittieria 2: 15. 1969. Type: Venezuela, Edo. Bolívar, near Guayaraca, Steyermark 94203 (holotype VEN, isotype NY).
Qualea apodocarpa Steyerm., Bol. Soc. Venez. Ci. Nat. 26: 473. 1966. – *Ruizterania apodocarpa* (Steyerm.) Marc.-Berti, Pittieria 2: 10. 1969. Type: Venezuela, Edo. Bolívar, Ichun R., Steyermark 90467 (holotype VEN).

Shrub or tree to 30 m high, stem 100 cm diam. Younger branchlets ferruginous-pubescent, older branchlets pubescent, cortex persistent to irregularly deciduous. Vegetative buds obtuse, 0.9-2.3 mm long, densely ferruginous-pubescent. Leaves opposite; stipular glands subprominent, ovate-subrounded, concave, 1.5-1.8 x 1.1-1.3 mm; petiole 3.5-6.5 mm long, glabrous to ferruginous-pubescent; blade obovate, lanceolate, oblong to elliptic, 3.2-9 x 1.2-4 cm, except primary vein pubescent, glabrous to ferruginous-pubescent below, margin slightly revolute, apex truncate-retuse to obtuse-retuse with mucro ca. 1 mm long, base obtuse to subrounded, sometimes rounded, glabrous above; primary vein impressed, glabrous to pubescent above, prominent, 2-winged, glabrous to densely pubescent below, wings glabrous to pubescent, ca. 0.2 mm wide, secondary veins to 35 per cm, slightly prominent on both surfaces, veinlets present towards margin, forming a network with long areolae, submarginal vein at 0.6 mm from margin. Panicle 5-15 cm long; cincinni subsessile, 2-flowered; pedicels 0.5 to 15 mm long on one inflorescence, densely ferruginous-pubescent. Flower buds densely ferruginous, generally obtuse at apex, ca. 3.5 mm wide above the smaller sepals; spurred sepal, including calyx tube, 12-18(-20) mm long at anthesis, densely ferruginous-pubescent outside, spur 4.0-9.5 x 1.7-3.5 mm, obtuse at apex, smaller sepals unequal, 3.5-8.5 mm long, acute at apex; petal white, spotted with orange, orbicular, ca. 18 mm long; staminal filament 7-14 mm long, anther (2.7-)5.0-9.0(-10.0) mm long, both thecae pubescent; ovary densely ferruginous-pubescent, style pubescent on the basal 1/5 part to almost completely. Fruit 1.5-4.5 cm long.

Distribution: Venezuela, Guyana and Brazil; ca. 54 collections studied (GU: 4).

Specimens examined: Guyana: Pakaraima Mts., Maguire 32251 (NY, U); Upper Mazaruni R., Pinkus 240; Tillett et al. 45030 (NY); Kamoa R., Jansen-Jacobs et al. 1700 (B, U).

Phenology: Flowering and fruiting almost during the whole year.

3. **Ruizterania rigida** (Stafleu) Marc.-Berti, Pittieria 2: 14. 1969. – *Qualea rigida* Stafleu, Acta Bot. Néerl. 2 (2): 162, fig. 7. 1953. Type: Venezuela, Edo. Bolívar, Caroní R., Cardona 1762 (holotype US, isotypes VEN, NY).

Qualea muelleriana M.R.Schomb., Reisen in Brit.-Guiana 3: 1099. 1848 (nomen nudum).

Shrub or tree to 8 m high. Younger branchlets glabrous to pubescent, cortex persistent, older branchlets generally glabrous, cortex cracked to deciduous. Vegetative buds acuminate, 1.0-1.5 mm long, glabrous. Leaves opposite; stipular glands slightly prominent, elliptic to subrounded 1.4-2.4 x 1.1-1.9 mm; petiole 3-6 mm long, pubescent; blade oblong-elliptic to ovate-lanceolate, 3-11 x 1.3-4 cm, margin plane, sometimes slightly revolute, apex acute to rounded with mucro ca. 1 mm long, base truncate, rounded to cordate at same branch, glabrous above, glabrous to glabrescent below; primary vein glabrous to glabrescent, plane to subplane above, prominent, pubescent and 2-winged below, wings ca. 0.25 mm wide, ciliate, secondary veins ca. 27 per cm, main secondary veins 10-12 per cm, subplane above, slightly prominent below, submarginal vein at 0.5-0.8 mm from margin. Panicle 9-1.25 cm long; cincinni 2-flowered; pedicels 7.0-9.5 mm long, ferruginous-pubescent. Flower buds ferruginous-pubescent outside; spurred sepal, including calyx tube, 20-26 mm long at anthesis, spur 7.5-8.0 x 2.3-2.8 mm, slightly incurved, smaller sepals unequal, 3.5-9.0 mm long; petal white, spotted with reddish yellow, 35-42 x 30-40 mm, glabrous, rudimentary petals 2-3 mm long; staminal filament 7.5-14.0 mm long, glabrous, anther 9-13 mm long, both thecae pubescent with erect, long hairs (± 0.8 mm long); staminodes 0.70 x 0.15 mm, glabrous; ovary ferruginous-pubescent, style pubescent on the basal 1/5. Fruit verruculose, ca. 2.5 cm long.

Distribution: Venezuela and Guyana; 11 collections studied (GU: 5).

Specimens examined: Guyana: Utshi R., CLAL 00215; Annaway valley, Rob. Schomburgk prob. I Add. 19; Upper Cujang R., Rich. Schomburgk 1537 (K); Upper Mazaruni R., Tillett & Tillett 45727 (NY), 45837 (NY, U).

Phenology: Flowering from June to January; fruiting from October to March.

4. **SALVERTIA** A.St.-Hil., Mém. Mus. Hist. Nat. 6: 259. 1820.
Type: Salvertia convallariodora A.St.-Hil.

Trees or shrubs; indument of simple hairs. Leaves in whorls; stipules deciduous; secondary veins 9-13 per side, ending in the margin or anastomosing irregularly. Inflorescences panicles of few-flowered cincinni. Flowers perigynous; sepals inserted on a very short calyx tube or hypanthium, subequal, spurred sepal not convolute, spur exserted in bud; petals 5; stamen 1, anther innate, connective not produced; staminodes 2; ovary 3-locular, with axile placentation and 2 ovules per locule, style club-shaped, stigma lateral. Capsules 3-locular, 1 winged seed per locule.

Distribution: Monotypic genus from Suriname, Brazil and Bolivia.

1. **Salvertia convallariodora** A.St.-Hil., (as 'convallariaeodora') Mém. Mus. Hist. Nat. 6: 266. 1820. Type: Brazil, Minas Gerais, St. Hilaire s.n. (holotype P, isotypes K, NY). – Plate 4 A-D.

   Salvertia thyrsiflora Pohl, Pl. Bras. Icon. Descr. 2: 16, t. 110. 1831. Type: Brazil, Goiás, Pohl s.n. (BR, M).

Tree or shrub to 3-8 m tall, stem up to 30 cm diam. Younger branchlets robust, pubescent, cortex persistent, cracked on older branchlets. Leaves up to 8 per whorl; stipules deciduous; petiole 15-40 mm long, glabrous to glabrescent; blade generally obovate, 10-27 x 6-17 cm, margin plane, apex rounded-retuse to emarginate, base acute, glabrous to glabrescent on both surfaces; primary vein almost plane above, sharply prominent below, main secondary veins 9-13 per side, 1-3 cm apart, prominent above, less so below, angle with primary vein ca. 50°, veinlets forming a network slightly prominent on both surfaces. Panicle terminal, up to 45 cm long; cincinni 2-3-flowered, ca. 9 cm long; peduncle of cincinni 20 - 25 mm long, densely pubescent; pedicels 10-20 mm long, densely pubescent. Flowers with agreeable fragrance; calyx greenish-yellow, spurred sepal, including calyx tube, 35-38 mm long at anthesis, spur green, 20-25 mm long and ca. 2.8 mm wide, at angle of 10°-90° with pedicel; petals yellow, subequal, ca. as long as stamen, glabrous; stamen glabrous, anther white with red spots, ca. 7 times longer than filament; staminodes glabrous, blade ovate to deltoid, 25-32 mm long and wide, filament 0.5-1.2 x 0.8-1.2 mm; ovary pubescent, style green, glabrous, stigma yellow, lateral, ca. 9.0 x 1.8 mm. Capsule ca. 50 mm long; seeds ca. 4 x 1 cm.

Distribution: See under the genus (SU: 3).

Specimens examined: Suriname: Sipaliwini savanna, Rombouts 275 (U), Oldenburger et al. 229 (U), van Donselaar 3607 (U).

Phenology: Flowering almost every month; fruiting from March to November.

5. **VOCHYSIA** Aubl., Hist. Pl. Guiane 1: 18. 1775; Poir., Encycl. 8: 681. 1808. – *Salmonia* Scopoli, Introd. 209. 1777. – *Vochya* Vell. ex Vandelli, Fl. Lusit. 1. 1788. – *Vochisia* Juss., Gen. Pl. 424. 1789. – *Cucullaria* Schreb., Gen. Pl. 6. 1789.
   Type: Vochysia guianensis Aubl.

   Strukeria Vell., Fl. Flum. 8. 1825.
   Type: Strukeria oppugnata Vell.

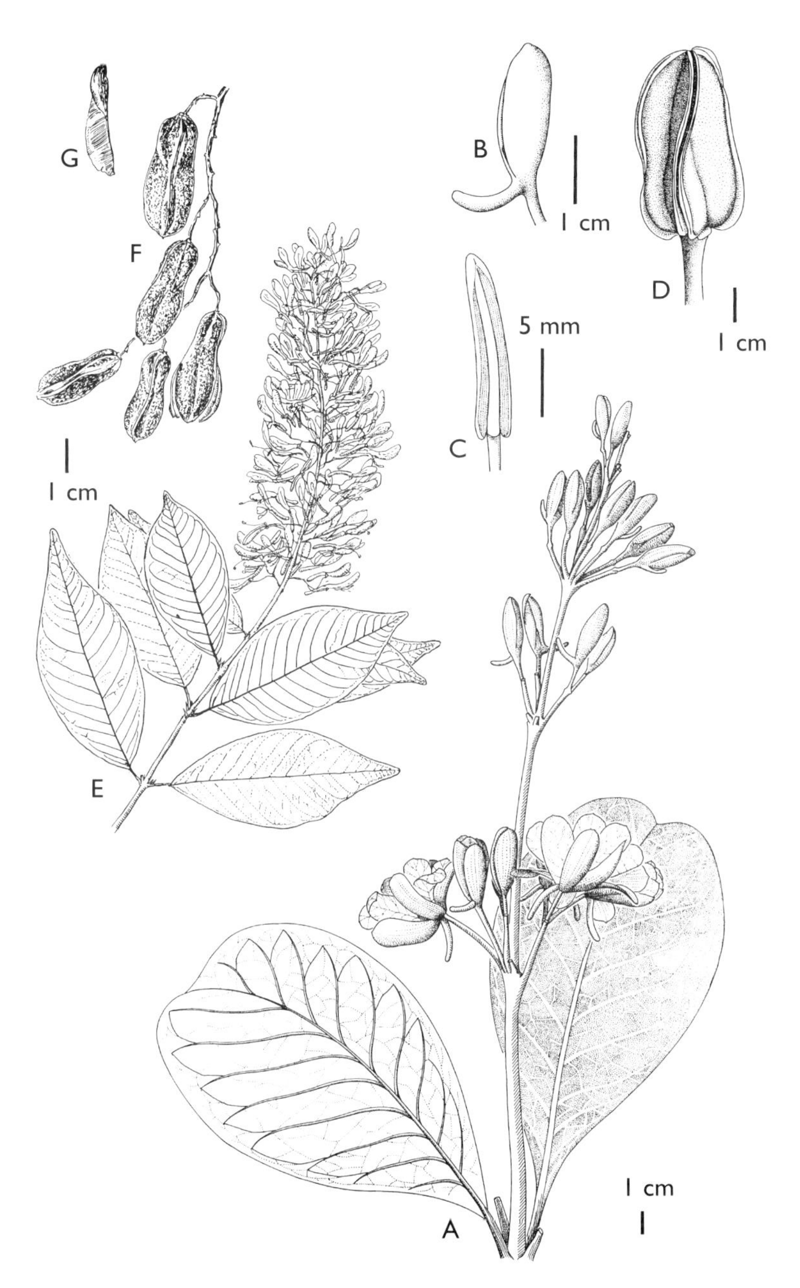

Plate 4. *Salvertia convallariodora* A.St.-Hil. A, flowering branch; B, flower bud just before opening; C, stamen; D, open fruit; *Vochysia tomentosa* (G.Mey.) DC. E, flowering branch; F, infructescence; G, seed. (A-D by G. Hintz; E-G by W.H.A. Hekking).

Trees, shrubs or rarely undershrubs; indument of simple or malpighioid (2-branched) hairs. Leaves opposite, in whorls, or sometimes scattered; stipules small, generally triangular; secondary veins 10-35 per side; marginal vein present or absent. Inflorescences panicles of 1-10-flowered cincinni. Flowers yellow, perigynous; sepals inserted on a very short calyx tube or hypanthium, posterior sepal spurred, 3-4 times longer than the others, convolute in bud, enveloping inner flower parts; petals free, imbricate, generally 3, sometimes 1, open in bud, stamen 1, in front of central petal, anther innate, connective cucullate; staminodes 2, opposite lateral petals; ovary 3-locular, 2 ovules per locule, stigma terminal to lateral. Capsules 3-locular, dehiscent; seeds winged, 1 per locule.

Distribution: Ca. 130 species from Mexico to Paraguay; 16 species occur in the Guianas.

Vernacular name: iteballi is used for most *Vochysia* species.

KEY TO THE SPECIES

1 Leaves in whorls of 3 to many · · · · · · · · · · 2
Leaves opposite · · · · · · · · · · 6

2 Blade densely appressed-pubescent with ferruginous, malpighioid hairs below · · · · · · · · · · *8. V. maguirei*
Blade glabrous to glabrescent, green to yellowish below · · · · · · 3

3 Leaves generally opposite, rarely in whorls of 3; spur longer than spurred sepal; ovary glabrous · · · · · · · · · · *7. V. guianensis*
Leaves in whorls of 3 to many; spur shorter or nearly as long as spurred sepal; ovary glabrous or pubescent · · · · · · · · · · 4

4 Leaves always in whorls of 3; ovary pubescent · · · · · · · · · · *13. V. surinamensis* var. *surinamensis*
Leaves in whorls of 4 to many, rarely of 3; ovary glabrous · · · · · · 5

5 Leaves in whorls of 4 rarely 3; anther glabrous · · · · · · *14. V. tetraphylla*
Leaves in whorls of 4 to many; anther ciliate at margin · · · · · · · · · · *16. V. tucanorum* var. *tucanorum*

6 Spur longer than spurred sepal; leaves rarely in whorls of 3 · · · · · · · · · · *7. V. guianensis*
Spur shorter or nearly as long as spurred sepal; leaves always opposite · · 7

7 Spur strongly incurved · · · · · · · · · · 8
Spur straight to slightly curved · · · · · · · · · · 9

8 Blade glabrous on both surfaces; spurred sepal, including calyx tube, 15-17 mm long at anthesis · 3. *V. crassifolia*
Blade densely malpighioid pubescent below, mainly along veins and veinlets; spurred sepal, including calyx tube, 7-10 mm long at anthesis · 5. *V. ferruginea*

9 Spur inflated, nearly twice as long as wide; anther about as long as staminal filament · 4. *V. densiflora*
Spur 4-7 times longer than wide; anther twice or more times longer than staminal filament · 10

10 Petal 1 · 11
Petals 3 · 12

11 Leaf glabrous below, except primary vein; spurred sepal, including calyx tube, 13-14 mm long at anthesis; anther ca. twice longer than filament; staminodes glabrous · 11. *V. schomburgkii*
Leaf densely appressed-pubescent with ferruginous, malpighioid hairs below; spurred sepal, including calyx tube, ca. 20 mm long at anthesis; anther 5-6 times longer than filament; staminodes ciliate · · · 9. *V. neyratii*

12 Secondary veins ca. 22 per side; stamen glabrous · · · · · · · 6. *V. glaberrima*
Secondary veins 9-13 or 30-35 per side; stamen pubescent dorsally and/or on the concave surface of the connective · 13

13 Blade densely pubescent below; petals pubescent dorsally · · · · · · · · · · · 14
Blade glabrous or glabrescent below; petals glabrous or ciliate at apex · · · 16

14 Secondary veins 30-35 per side, at 2-7 mm distance · · · · · · 10. *V. sabatieri*
Secondary veins 9-13 per side, at 5-11 mm distance · · · · · · · · · · · · · · · 15

15 Petiole 10-17 mm long; stigma terminal · · · · · · · · · · · · · · · · · 2. *V. costata*
Petiole 3-8 mm long; stigma lateral · · · · · · · · · · · · · · · · · · 15. *V. tomentosa*

16 Petiole 5-12 mm long; blade 6.5-8.5 x 3.0-4.2 cm, apex obtuse to acuminate; anther completely pubescent on both sides, 7-8 times longer than filament · 1. *V. cayennensis*
Petiole 20-30 mm long; blade 9-15 x 5-7 cm, apex rounded to truncate; anther pubescent only on the concave surface, ca. 3 times longer than filament · 12. *V. speciosa*

1. **Vochysia cayennensis** Warm. in Mart., Fl. Bras. 13(2): 80. 1875. Type: French Guiana, Martin s.n. (holotype P, isotypes B-W, BM, K, NY).

Tree 30-35 m high, trunk up to 80 cm diam. Leaves opposite; stipules nearly 1 mm long; petiole 5-12 mm long; blade generally elliptic, 6.5-8.5

x 3-4.2 cm, margin slightly revolute, apex obtuse to acuminate, base cuneate, glabrous on both surfaces; primary vein prominent below, secondary veins 10-12 per side, angle with primary vein ca. 60°, scarcely prominent below, submarginal vein absent. Inflorescence terminal and axillary, up to 22 cm long; cincinni 1(-2)-flowered; peduncle and pedicel together 7-12 mm long, sparsely pubescent. Spurred sepal, including calyx tube, 15.5-19.0 mm long at anthesis, glabrous to glabrescent, spur slightly recurved, 5-7 mm long, at an angle of 25°-30° with pedicel, smaller sepals subequal, 3.0-3.5 mm long; petals 3, glabrous, central petal ca. 4-5 times shorter than stamen and slightly longer than lateral petals; staminal filament 7-8 times shorter than anther, glabrous, anther pubescent on both sides; staminodes triangular, ca. 1.0 x 0.7 mm, ciliate along margin; ovary glabrous, stigma partially terminal, ca. 0.5 x 0.5 mm.

Distribution: French Guiana; 9 collections studied.

Selected specimens: French Guiana: Camopi R., Oldeman & Sastre 56 (P), 213 (P).

Phenology: Flowering from December to February.

Vernacular names: French Guiana: grignon fou, kouali Sainte Marie.

Note: Desfontaines s.n., studied in herbarium Willdenow (B), is probably a duplicate of Martin s.n.; Stafleu cited the C voucher as cotype of *V. cayennensis*.

2. **Vochysia costata** Warm. in Mart., Fl. Bras. 13(2): 100. 1875. Type: Guyana, Rich. Schomburgk 974 (lectotype P, isolectotypes BM, G).

Tree up to 30 m high. Older branchlets quadrangular, pubescent, cortex persistent. Leaves opposite; stipules triangular, 2.5-3.0 mm long, acuminate, pubescent; petiole 10-17 mm long, pubescent; blade elliptic, 12-14 x 4-6 cm, margin plane, undulate, apex acuminate, base acute, glabrous above except primary vein pubescent at least towards base, densely ferruginous-pubescent with malpighioid appressed hairs below; primary vein impressed above, prominent below; main secondary veins 12-13 per side, at 6-11 mm distance, plane or subplane above, subprominent below, slightly arcuate, angle with primary vein 45°-50°, veinlets inconspicuous to plane above, scarcely prominent below, submarginal vein present at least towards apical third of blade. Inflorescence terminal and axillary, 19-24 cm long; cincinni

2-3-flowered; peduncle 6.8-8.0 mm long, pubescent; pedicels 4.6-7.0 mm long, pubescent. Spurred sepal, including calyx tube, 12.5-14.5 mm long at anthesis, sparsely pubescent, spur 7.0-9.5 mm long, 1.9-2.5 mm wide towards base and 0.8 mm wide near apical 1/3, subacute at apex, incurved at an angle of 90°-120° with spurred sepal, smaller sepals subequal, 3.2-3.5 mm long, ciliate; petals 3, unequal, central petal as long as stamen, pubescent dorsally, obtuse at apex; lateral petals pubescent on back, approximately 1/3 shorter than central petal; anther pubescent on internal surface and on margin, approximately twice longer than glabrous filament; staminodes glabrous, 0.9-1.0 x 0.3-0.5 mm; ovary glabrous, stigma terminal.

Distribution: Venezuela, Guyana and Brazil; 4 collections studied, from Guyana only the type.

Phenology: Flowering in June; fruiting in March.

Vernacular names: Suriname: asjiwa (Paramacca). Guyana: iteballi (Arawak).

3. **Vochysia crassifolia** Warm. in Mart., Fl. Bras. 13(2): 77. 1875. Type: Guyana, Roraima, Rob. Schomburgk 585 (holotype K, isotype G).

   Vochysia curvata Klotzsch in M.R.Schomb., Reisen in Brit.-Guiana 3: 1099. 1848 (nomen nudum).

Tree up to 20 m high, trunk 10-47 cm diam. Younger branchlets angled, glabrous to glabrescent, older branchlets generally greyish, cortex deciduous. Leaves opposite; stipules 1.6-2.2 mm long, pubescent; petiole 10-17 mm long, glabrous to glabrescent, blackish; blade elliptic to elliptic-lanceolate, (6.5-)8-18 x (3.4-)4-8 cm, margin plane to slightly revolute, apex acuminate, base subrounded to rounded, glabrous on both surfaces; primary vein subimpressed to plane above, prominent below, main secondary veins 14-17 per side, plane to subplane above, subplane below, angle with primary vein 70°-75°, arcuately anastomosing at 4.0-6.5 mm from margin, minor secondary veins 1-2 in between each pair of main ones and shorter, veinlets generally inconspicuous below. Inflorescence terminal and axillary, 12-23 cm long; cincinni 1-2-flowered, 2-2.5 cm long; peduncle 3.0-5.7 mm long, pubescent; pedicels 2.5-3.0 mm long, pubescent. Spurred sepal, including calyx tube, (6-8-)15-17 mm long at anthesis, appressed-pubescent, spur pubescent, strongly incurved, at an angle of 20°-60° with pedicel, smaller sepals 3.5-5.0 mm long, ciliate, pubescent; petals 3, central petal ca. half as long as stamen, with very narrow line of hairs on back in basal dorsal half; lateral petals ciliate, shorter; stamen glabrous; staminodes oblanceolate, glabrous, ca. 1.2 x 0.4 mm; ovary glabrous, stigma terminal, 0.4-0.6 x 0.4-0.6 mm. Fruit ca. 3 cm long.

Distribution: Venezuela, Guyana and Brazil; 17 collections studied, 2 from Guyana: Roraima, Schomburgk II 585/Rich. 964 (K); Rupununi R., FD 2246 (K).

Phenology: Flowering from November to January; fruiting in April.

Vernacular name: Guyana: iteballi (Arawak).

4. **Vochysia densiflora** Spruce ex Warm. in Mart., Fl. Bras. 13(2): 101, t. 19. 1875. Type: Brazil, Amazonas, Spruce 2627 (holotype K, isotypes BM, BR, C, G, GH, NY, OXF). – Plate 5 D-F.

Tree up to 30 m high. Branchlets robust, angled, densely appressed-pubescent with malpighioid hairs. Leaves opposite; stipules ca. 1 mm long; petiole 10-17 mm long, densely appressed-pubescent; blade elliptic to ovate-elliptic, 8-15 x 3.8-7.5 cm (according to Stafleu up to 25 x 9 cm), margin plane, apex rounded to obtuse-retuse, base obtuse, glabrous above, densely ferruginous to greyish appressed-pubescent with malpighioid hairs all over below; primary vein impressed above, prominent below, main secondary veins ca. 15 per side, at 3-13 mm distance, angle with primary vein 50°-60°, minor secondary veins 1-2 between each pair of main ones and shorter and thinner, veinlets numerous, arcuate or S-shaped, generally scalariform between secondary veins, submarginal vein crenate at 0.9-1.5 mm from margin. Inflorescence terminal, densely flowered, ca. 10 cm long; cincinni 2-4-flowered, up to 4.7 cm long; pedicel 9.0-12.7 mm long. Spurred sepal, including calyx tube, 20-22 mm long at anthesis, spur 6.6-8.0 x 4.0-5.0 mm, inflated, oriented along pedicel, wider than flower-bud, smaller sepals subequal, more or less 5 mm long; petals 3, appressed-pubescent dorsally, central petal ca. 3 times shorter than stamen, but 4 times longer than lateral petals; stamen, except thecae, totally appressed-pubescent, filament as long as anther; staminodes nearly 1.4 mm long, pubescent with hairs up to 0.9 mm long; ovary glabrous, stigma lateral, ca. 0.9 mm long. Capsule 4-4.5 x 1.5-2 cm.

Distribution: Colombia, the Guianas and Brazil; 20 collections studied (GU: 2; SU: 8; FG: 7).

Selected specimens: Guyana: Moraballi Cr., FD 3176 (K); Demerara R., FD 3089 (K). Suriname: Moengo tapoe line to Nassau

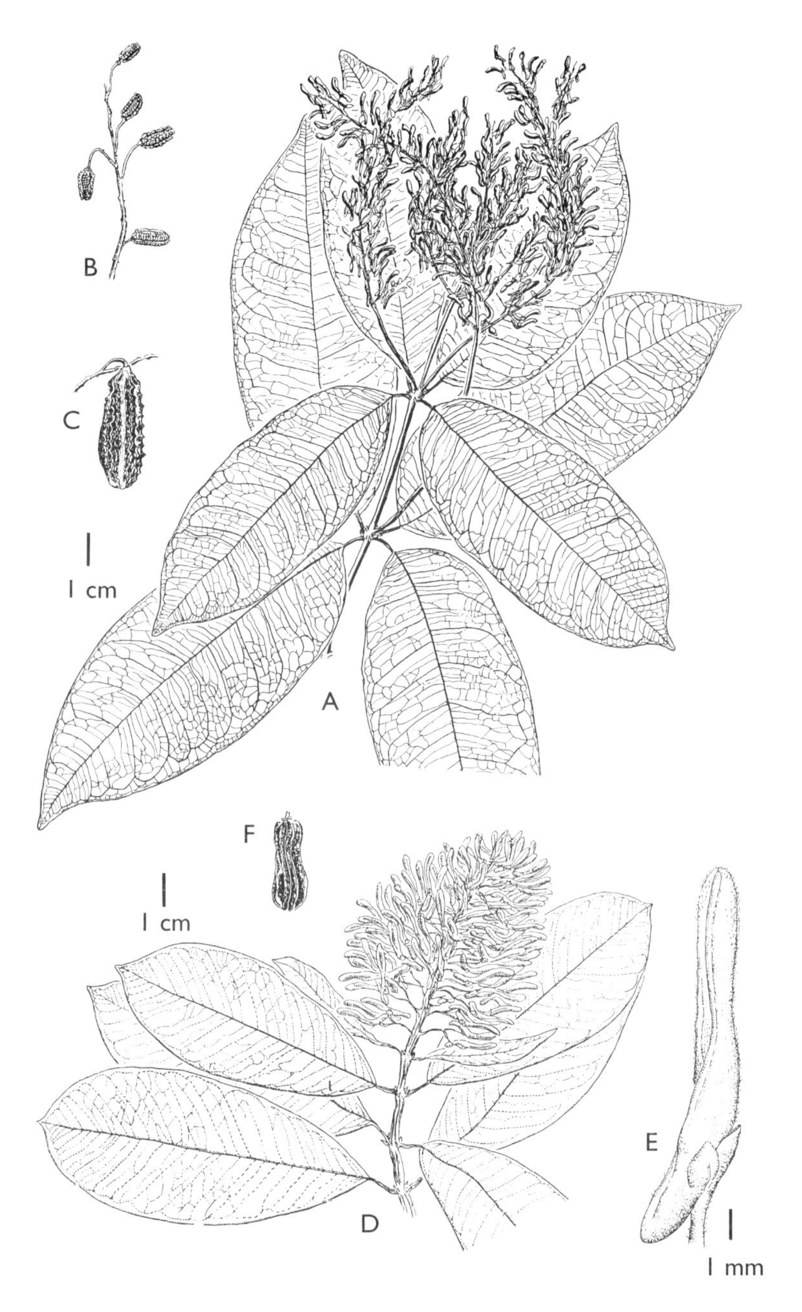

Plate 5. *Vochysia tetraphylla* (G.Mey.) DC. A, flowering branch; B, infructescence; C, open fruit; *Vochysia densiflora* Spruce ex Warm. D, flowering branch; E, flower bud just before opening; F, fruit. (Drawing by W.H.A. Hekking).

Mts., Lanjouw & Lindeman 539 (U), 2854 (U); Zanderij I, Woodherbarium 111 (U), BW 6052 (U). French Guiana: BAFOG 7190 (P); 7505 (P).

Phenology: Flowering from November to April; fruiting from February to May.

Vernacular names: Suriname: aprakwari (Sranan); appelkwari (Suriname-Dutch); iteballi, iteballi hariraru (Arowak); asjiwa, wanakwali (Paramacca). French Guiana: wana-kouali.

5. **Vochysia ferruginea** Mart., Nov. Gen. Sp. Pl. 1: 151. t. 92. 1824. – *Cucullaria ferruginea* (Mart.) Spreng., Syst. Veg. 4. Cur. Post 9. 1827. – *Vochysia ferruginea* (Mart.) Standl., N. Amer. Fl. 25: 302. 1924. Type: Brazil, Rio Negro, Martius s.n. (holotype M, isotypes K, L).

Tree up to 25 m high, trunk 20-35 cm diam. Younger branchlets densely ferruginous-pubescent, older branchlets subquadrangular to subterete, ferruginous-pubescent. Leaves opposite; stipules triangular-lanceolate, ca. 2 x 1 mm, densely pubescent; petiole 8-10 mm long, densely pubescent; blade elliptic, oblanceolate to lanceolate, 6.5-16.5 x 2.1-5.5 cm, margin subrevolute, apex acuminate, acumen 5-25 mm long, base obtuse, glabrous above, densely pubescent below, mainly along the veins and veinlets with 2-branched hairs (one branch erect or undulate, one short or inconspicuous); primary vein impressed above, prominent below, secondary veins 12-13 per side, at 3.5-8.5 mm distance, subplane above, subprominent below, angle with primary vein 30°-40°, veinlets numerous, marginal vein present near apical fourth of blade. Inflorescence terminal and axillary, 6-13 cm long; cincinni 1-5-flowered; peduncle 2.2-4.5 mm long; pedicels 4.0-4.5 mm long. Spurred sepal, including calyx tube, 7-10 mm long at anthesis, spur slightly shorter than spurred sepal, strongly incurved, smaller sepals 2.5-3.1 mm long; petals 3, pubescent dorsally, central petal ciliate at apex, as long or slightly shorter than stamen, lateral petals slightly shorter than central petal; staminal filament 1.7-2.4 mm long, glabrous to glabrescent dorsally, but pubescent in central portion, anther 5.0-5.5 mm long, glabrescent on lateral parts of external surface, but pubescent on inner surface, at least on basal half; staminodes oblong to oblanceolate, 0.7 x 0.4 mm, glabrous; ovary glabrous, stigma lateral, ca. 0.5 x 0.3 mm. Capsule 2-2.5 x 0.7-1.1 cm.

Distribution: Nicaragua, Costa Rica, Panama, Colombia, Venezuela, Guyana, Peru and Brazil; 27 collections studied (GU: 5).

Selected specimens: Guyana: Parabara savanna, FD 7643 (K); Rupununi R., A.C. Smith 2511; Essequibo R., A.C. Smith 2807.

Phenology: Flowering from December to November; fruiting from March to December.

Vernacular name: Guyana: iteballi (Arawak).

6. **Vochysia glaberrima** Warm. in Mart., Fl. Bras. 13(2): 78. 1875. Type: Guyana, Roraima, Rob. Schomburgk 642 (holotype K, isotypes BM, G, GH, L, OXF, P, W).

Vochysia lucida Klotzsch in M.R.Schomb., Reisen in Brit.-Guiana 3: 1099. 1848 (nomen nudum).

Tree 8-35 m high, trunk 20-80 cm diam. Younger branchlets blackish, sparsely pubescent, older branchlets reddish, cortex deciduous. Leaves opposite; stipules deltoid, 1.4 mm long, acute, pubescent; petiole 12-25 mm long, glabrous or pubescent; blade lanceolate to elliptic, 8-12.7 x 2.2-4.5 cm, margin subrevolute, apex obtuse to obtuse-acuminate, base cuneate, revolute, glabrous on both surfaces; primary vein plane to subimpressed above, prominent below, secondary veins ca. 22 per side, plane to subplane above, slightly prominent below, angle with primary vein 55°-60°, anastomosing 3-5 mm from margin, veinlets plane above, subplane to plane below. Inflorescence terminal and axillary, 14-28 cm long; cincinni 2-3-flowered, 1.9-2.6 cm long; peduncle ca. 6 mm long, glabrous to sparsely pubescent; pedicels 4.5-7.0 mm long, glabrous to sparsely pubescent. Spurred sepal, including calyx tube, 10.5-13.5 mm long at anthesis, glabrous, spur 6.5-10.0 x 1.3-1.4 mm, straight to slightly incurved, at an angle of 20°-85° with pedicel, spur plus spurred sepal arched to U-shape, smaller sepals ca. 3.4 mm long, ciliate; petals 3, ciliate at apex, central petal approximately 1/3 shorter than stamen and slightly longer than lateral ones; stamen glabrous; staminodes glabrous, oblanceolate, 0.7 x 0.25 mm; ovary glabrous, stigma lateral, 0.7-0.8 x 0.4-0.7 mm.

Distribution: Venezuela, Guyana, Suriname and Brazil; 8 collections studied (GU: 4; SU: 1).

Specimens examined: Guyana: Rupununi R., A.C. Smith 2258, 2425, Jansen-Jacobs et al. 1623 (U). Suriname: Maratakka R., LBB 10821 (U).

Phenology: Flowering from August to November.

Vernacular names: Guyana: iteballi (Arawak); deokunud (Wapisiana).

7. **Vochysia guianensis** Aubl., Hist. Pl. Guiane 1: 18, t. 6. 1775. (as 'Vochy'). – *Vochisia guianensis* (Aubl.) Lam., Tabl. Encycl. 3(1): 35, t. 2. 1791. – *Cucullaria excelsa* Willd., Spec. Pl. 1: 17. 1797 (non Vahl). Type: French Guiana, Aublet s.n. (holotype BM).

Vochysia melinonii Beckmann in Engl., Bot. Jahrb. 40: 280. 1908. Type: French Guiana, Maroni, Mélinon s.n. (1863) (holotype P, isotypes BM, K, NY).
Vochysia paraensis Huber ex Ducke, Arch. Jard. Bot. Rio de Janeiro 1: 44. 1915. Type: Brazil, Pará, Huber MG 538 (holotype BM, isotypes G, P, US).

Tree 25-35 m high, stem ca. 50 cm diam. Younger branchlets blackish, sparsely pubescent, cortex persistent, older branchlets with deciduous cortex. Leaves opposite, rarely in whorls of 3; stipules ca. 1.6 mm long; petiole 10-20 mm long; blade oblong, elliptic to ovate, 7-15.5 x 3.5-5.7 cm, margin subrevolute, apex obtuse-retuse to rounded-retuse, base cuneate and subrevolute, glabrous on both surfaces; primary vein plane to subplane above, prominent below, main secondary veins 13-16 per side, angle with primary vein 60°-65°, main secondary veins bifurcating at 3-7 mm from margin and connected with adjacent ones, minor secondary veins 2-3 between each pair of main ones and shorter, forming a more or less homogenous network, visible on both surfaces, more conspicuous below. Inflorescence terminal and axillary, up to 20 cm long; cincinni 3-5-flowered, 1.9-2.3 cm long; peduncle 5-10 mm long, glabrous; pedicels up to 6 mm long, glabrous. Spurred sepal, including calyx tube, 6-8 mm long at anthesis, glabrous to glabrescent, spur 8-10 x 1.6 mm, incurved, smaller sepals 2.2-3.0 mm long, pubescent or only ciliate; petals 3, glabrous dorsally, ciliate at apex, central petal nearly 1/3 shorter than stamen, lateral petals 1/4 shorter than central petal; staminal filament as long as anther, glabrous, anther glabrous or with 1-5 hairs dorsally; staminodes triangular or rhomboid, glabrous, 0.7 x 0.5-0.6 mm; ovary glabrous, stigma lateral, 1.0-2.0 x 0.5 mm, more or less elliptic. Fruit up to 5.2 x 2.5 cm.

Distribution: Suriname, French Guiana and Brazil; 34 collections studied (SU: 13; FG: 18).

Selected specimens: Suriname: Jodensavanne-Mapane Cr. area, Lindeman 5208 (U); 5243 (U). French Guiana: St. Laurent, BAFOG 7066, 7105, 7115, 7165 (P).

Phenology: Flowering from October to December; fruiting from February to April.

V e r n a c u l a r n a m e s : Suriname: gwanakwari, wiswiskwari (Sranan); iteballi kuleru (Arowak); asjiwa, moetende (Paramacca); rode kwari (Suriname-Dutch); wosi-wosi (Carib). French Guiana: bois cruzeau, moutendé, moutendé-kouali.

8. **Vochysia maguirei** Marc.-Berti, Pittieria 9: 29, fig. s.n. 1981. Type: Guyana, Mazaruni R., Maguire 32645 (holotype MER, isotypes NY, VEN).

Tree 17 m high; indument mainly of malpighioid hairs. Younger branchlets densely ferruginous-pubescent, older branchlets quadrangular with concave sides, densely or partially pubescent, glabrescent. Leaves in whorls of 4-5; stipules triangular, ca. 3 mm long; petiole 5.5-8 mm long; blade spatulate, 24-35 x 6-9 cm; primary vein densely appressed ferruginous-pubescent, margin plane, apex acuminate, base attenuate towards petiole, sparsely pubescent above, densely so below; primary vein impressed above, prominent and robust below, main secondary veins 24-29 per side, 7-17 mm apart, straight or almost so, angle with primary vein 50°-55°, veinlets numerous, submarginal vein at 1-3 mm from margin. Inflorescence ca. 17 cm long; cincinni 5-6-flowered. Spurred sepal, including calyx tube, 12-14 mm long at anthesis, spur 7.5-8.0 x ca. 4 mm, straight to substraight, at an acute angle with spurred sepal, spur plus spurred sepal V-shaped, smaller sepals subequal, ca. 4 mm long; petals 3, unequal, central petal slightly shorter than stamen, densely ferruginous appressed-pubescent dorsally except ciliate apex, lateral petals half as long as petal, sparsely pubescent dorsally at base, glabrous elsewhere; staminal filament glabrous, anther ca. 3 times longer than the filament, obtuse at apex, pubescent on both sides; ovary glabrous, stigma lateral-terminal, 0.3 mm long, 0.5 mm wide.

D i s t r i b u t i o n : Known from the type collection and Cuyuni-Mazaruni Region, along Koatse R., ca. 20 km W of Pang R., Pipoly et al. 10578 (B).

P h e n o l o g y : Flowering in November.

9. **Vochysia neyratii** Normand, Adansonia ser. 2, 17(1): 11, pl. 1. 1977. Type: French Guiana, Comté R., Petrov 185 (holotype P).

Tree to 35 m high, trunk up to 1 m diam. Very young branchlets densely malpighioid ferruginous-pubescent, cortex of younger and older branchlets deciduous. Leaves opposite; stipules triangular, ca. 3.5 x 1.5 mm, acuminate; petiole 10-15 mm long; blade narrowly elliptic, 11-17 x

3-6 cm, margin not revolute, apex acuminate, 10-15 mm long, base obtuse; densely appressed-pubescent with ferruginous malpighioid hairs below, glabrous above; primary vein impressed above, prominent below, main secondary veins 25-30 per side, arcuate, at angle of 50°-60° with primary vein, minor secondary veins between each pair of main ones, submarginal vein at 5-10 mm from margin. Inflorescence 20-25 cm long; cincinni 1-2-flowered; peduncle ca. 0.7 cm long, ferruginous-pubescent; pedicels nearly 1.2 mm long, ferruginous-pubescent. Spurred sepal, including calyx tube, ca. 20 mm long at anthesis, ferruginous-pubescent, spur slightly incurved, nearly 7.5 mm long, at an angle of 35°-40° with pedicel, smaller sepals subequal, ca. 3.5 mm; only central petal present, densely appressed ferruginous-pubescent dorsally; stamen 3-4 times longer than petal, anther 5-6 times longer than filament, obtuse at apex, glabrous outside, densely ferruginous-pubescent on margin and inside between thecae; staminodes ciliate, 1.0 x 0.8 mm; ovary glabrous, stigma terminal, subcapitate, 0.3 x 0.7 mm. Capsule 3-4 cm long.

Distribution: Only known from French Guiana; 20 collections studied.

Selected specimens: French Guiana: Comté R., Petrov 168, 184 (P); Upper Oyapock, Sastre 4415 (P, U), 4500 (P).

Phenology: Flowering from January to March; fruiting from February to April.

Vernacular names: French Guiana: achiwat, kouali-neyrat, kwali-pita, kwali-sili.

Note: Normand describes 3 petals in the original description of the species.

10. **Vochysia sabatieri** Marc.-Berti, Pittieria 18: 9. 1989. Type: French Guiana, Sabatier 1170 (holotype NY, isotype CAY).

Small tree. Younger branchlets quadrangular densely appressed ferruginous-pubescent. Leaves opposite; stipules pubescent, triangular, ca. 5 mm x 3.0 mm, stipular ridge present; petiole 12-16 mm long, densely appressed-pubescent; blade elliptic to subelliptic, 17-21 x 5-6 cm, apex acuminate, acumen ca. 10 mm long, base obtuse, glabrous above, except primary vein sparsely pubescent at least towards base, densely appressed malpighioid ferruginous-pubescent below; primary vein prominent below, impressed above, secondary veins 30-35 per side, at 2-7 mm distance, plane to subplane above, less prominent than primary vein below, veinlets numerous, forming a conspicuous network, submarginal vein at 0.5-1.0 mm from margin. Inflorescence terminal, ca. 20 cm long; cincinni 2-3-flowered. Spurred sepal, including calyx tube, ca. 16 mm long, spur ca. 9 x 2 mm, slightly curved, at an acute angle with pedicel, smaller sepals 2.7-3.2 mm long, ciliate; petals 3, central petal 1/6-1/8 shorter than the stamen, pubescent dorsally, lateral petals 1/3-1/4 shorter than central petal; staminal filament pubescent at least partially, anther pubescent, except for glabrous thecae, nearly 2.5 times longer than filament; staminodes ciliate, 1.6 x 0.8 mm without hairs; ovary glabrous, stigma lateral, 0.8 x 0.8 mm, slightly bilobed at base.

Distribution: French Guiana: known only from the type collection.

Phenology: Flowering in December.

11. **Vochysia schomburgkii** Warm. in Mart., Fl. Bras. 13(2): 78. 1875. Type: Guyana, Upper Demerara R., Rob. Schomburgk II 902/Rich. 1360 (holotype K, isotypes BM, G, P, W).

Small tree. Younger branchlets quadrangular, appressed-pubescent, older branchlets subterete, glabrous. Leaves opposite; stipules triangular, ca. 1 mm long; petiole ca. 10 mm long; blade 8-15 x 3.8-6.5 cm, except on primary vein sparsely pubescent, margin plane, apex acuminate, base obtuse, glabrous on both surfaces; secondary veins 10-12 per side, angle with primary vein ca. 70°, submarginal vein absent. Inflorescence terminal, ca. 7 cm long; cincinni (1-)2-3-flowered; peduncle 3-4 mm long; pedicels 4-7 mm long, pubescent. Spurred sepal, including calyx tube, 13-14 mm long at anthesis, spur straight or slightly curved, inflated towards the middle, ca. 7 mm long, at an angle of 90° with pedicel, smaller sepals ca. 2.5 mm long; only central petal present, densely pubescent dorsally; stamen ca. 2.5 x longer than petal, staminal filament glabrous, anther ca. twice longer than filament, glabrous dorsally, densely pubescent on margin and on inner surface; staminodes triangular, obtuse, glabrous, nearly 0.6 x 0.5 mm; ovary glabrous, stigma terminal, 0.3 x 0.5 mm.

Distribution: Guyana; 6 collections studied.

Selected specimen: Guyana: Makauria Cr., FD 3260 (K).

Phenology: Flowering in July and February.

Vernacular names: Guyana: iteballi (Arawak); tuacoo.

12. **Vochysia speciosa** Warm. in Mart., Fl. Bras. 13(2): 79. 1875. Type: French Guiana, Poiteau s.n. (holotype LE not seen, isotypes G, K, P, W).

Tree up to 45 m high. Young branchlets quadrangular, cortex persistent. Leaves opposite; stipules ca. 2 mm long; petiole 20-30 mm long; blade generally obovate, 9-15 x 5-7 cm, apex rounded to truncate, base obtuse, glabrous on both surfaces; primary vein prominent below, main secondary veins 10-15 per side, angle with primary vein 60°-80°, submarginal vein absent. Inflorescence terminal, ca. 25 cm long; cincinni 2-5-flowered, ca. 3 cm long; peduncle nearly 1 cm long, glabrous; pedicels 10-12 mm long, glabrous. Spurred sepal, including calyx tube, ca. 18 mm long at anthesis, spur 7-9 mm long, smaller sepals subequal, 2.8-3.2 mm long; petals 3, glabrous, ciliate at apex, central petal 1/2-1/3 shorter than stamen, lateral petals slightly shorter than central petal; anther nearly 3 times longer than filament, glabrous dorsally, pubescent on internal surface; staminodes glabrous, triangular, 0.8-1.0 x 0.4 mm; ovary glabrous, stigma subterminal, 0.3 x 0.6 mm.

Distribution: French Guiana; 8 collections studied.

Selected specimens: French Guiana: Kaw Mt., Cowan 38779; Roura Mt., Petrov 111, 112, 163 (P).

Phenology: Flowering in December; fruiting in January.

Vernacular names: French Guiana: kouali-rougier, wachi-wachi-kouali.

13. **Vochysia surinamensis** Stafleu, Rec. Trav. Bot. Néerl. 41: 439. 1948.

In the Guianas only: var. **surinamensis**. Type: Suriname, Brownsberg, BW 6915 (holotype U).

Tree 8-40 m tall, trunk up to 55 cm diameter. Younger branchlets angled, blackish, glabrous to glabrescent, cortex persistent, cortex of older branchlets deciduous. Leaves in whorls of 3; stipules triangular, ca. 1.0 x 0.7 mm, pilose; petiole 10-15 mm long, glabrous; blade obovate to elliptic-obovate, 5 x 2.5-13 x 5 cm, glabrous on both surfaces, margin subrevolute, apex retuse-obtuse to retuse-rounded, base cuneate, glabrous on both surfaces; primary vein impressed above, prominent below, main secondary veins ca. 24 per side, at 2-3 mm distance, slightly prominent to subplane on both surfaces, veinlets numerous, forming a clearly defined network on both surfaces, submarginal vein not evident. Inflorescence a terminal or axillary panicle, 20-25 cm long, axis pubescent; cincinni (1-)2-3-flowered; pedicels 4.0-4.5 mm long, densely pubescent. Calyx appressed-pubescent outside, spurred sepal, including calyx tube, 10-12 mm long at anthesis, spur sparsely pubescent, 7.5-9.0 mm long, inflated near base (ca. 2 mm wide) and narrowed near apex (ca. 1 mm wide), straight to slightly incurved, at an angle of 35°-60° with spurred sepal and 65°-70° with pedicel, smaller sepals subequal, ca. 2.7 mm long; petals 3, glabrous, central petal slightly longer than lateral petals and almost half as long as stamen; stamen glabrous, anther 3.5-4 x longer than filament; ovary pubescent, style pilose on lower 1.5-2.5 mm, stigma lateral-terminal, ca. 0.4 x 0.7 mm. Fruit 3.5-4.5 cm long.

Distribution: Venezuela, the Guianas and Brazil; 18 collections studied (GU: 3; SU: 6; FG: 4).

Selected specimens: Guyana: Arawai Cr., Fanshawe 3050 (K); Demerara R., FD 5218 (K). Suriname: Brownsberg, BW 3260, 6510, 6915 (U). French Guiana: BAFOG 309M, 7725 (P); Saül, Pelée Mt., Oldeman B-4005 (P).

Phenology: Flowering from May to December; fruiting from January to November.

Vernacular names: Guyana: hill iteballi, iteballi (Arawak). Suriname: asjiwa (Paramacca); iteballi (Arowak); kwari, wanakwari (Sranan). French Guiana: grignon fou, kouali, wana-kouali.

Note: *Vochysia surinamensis* Stafleu var. *inflata* Stafleu occurs on tepuis in estado Bolívar, Venezuela.

14. **Vochysia tetraphylla** (G.Mey.) DC., Prodr. 3: 27. 1828. – *Vochisia tetraphylla* (G.Mey.) Stone & Freeman, Timber Brit. Guiana 26.1914. – *Cucullaria tetraphylla* G.Mey. Prim. Fl. Esseq. 12. 1818. Type: Guyana, Rodschied s.n. (not seen). – Plate 5 A-C.

    Vochysia arcuata Garcke, Linnaea 22: 58. 1849. Type: Suriname, Kegel 684, (not seen).

Tree up to 40 m high, trunk up to 100 cm diameter. Younger branchlets, angled, glabrous, yellowish to blackish, older branchlets subtetragonous, glabrous to glabrescent, brownish to blackish, cortex generally persistent. Leaves in whorls of 4, rarely 3; stipules deltoid, 1.5-1.8 mm long, acuminate, pubescent; petiole 3-8.5 mm long, glabrous to glabrescent, blackish; blade elliptic, lanceolate to oblanceolate, 9-20 x 3-8.5 cm, margin plane to subplane, apex acuminate, base obtuse to rounded, glabrous above, glabrescent below; primary vein plane to subplane above, prominent below, main secondary veins 14-20 per side, plane to subplane above, scarcely prominent below, angle with primary vein 70°-80°, irregularly anastomosing at 3.5-10 mm from margin, minor secondary veins between each pair of main ones and shorter, veinlets numerous, plane above, subplane below. Inflorescence terminal and axillary, 10-22 cm long; cincinni generally 1-2-flowered, ca. 2 cm long; peduncle 3.0-3.5 mm long, sparsely pubescent; pedicels 6.5-9.0 mm long, sparsely pubescent. Spurred sepal, including calyx tube, 12.0-14.5 mm long at anthesis, spur subconic, straight to slightly recurved, 5.0-6.5 x 1.2-1.6 mm, at an angle of 0°-20° with pedicel, spur plus spurred sepal S-shaped, smaller sepals 2.5-3.0 mm long, sparsely pubescent, ciliate; petals 3, glabrous, central petal ciliate at apex, half as long as stamen, slightly longer than lateral petals; stamen glabrous, anther ca. twice longer than filament; staminodes subspatulate, 0.7 x 0.3 mm, glabrous; ovary glabrous, stigma terminal-lateral, 0.6 x 0.6 mm. Capsule 3.2 x 3.2 cm, rugulate.

Distribution: Venezuela, the Guianas and Brazil; ca. 70 collections studied (GU: 6; SU: 25; FG: 25).

Selected specimens: Guyana: Rupununi, Cook 244 (K, M); Kanuku Mts., Jansen-Jacobs et al. 680 (U). Suriname: Albina, Marowijne R., Lanjouw & Lindeman 281. French Guiana: St. Laurent, Portal Cr., Oldeman 2264 (P); between Bonidoro and Bolimon Fou, Sabatier 990 (P).

Phenology: Flowering from August to January; fruiting from March to August.

Vernacular names: Guyana: iteballi (Arawak); kuwariri (= kwariri = woto-kwari (woto = fish) (Carib); kwaru (Akawaio). Suriname: watrakwari (Sranan); iteballi unirefodikoro (Arowak); asjiwa, papakai-kwali (Paramacca). French Guiana: kouali de rivière.

15. **Vochysia tomentosa** (G.Mey.) DC., Prodr. 3: 26. 1928. – *Cucullaria excelsa* Vahl, Enum. Plant. 1: 4. 1804 (non Willd. 1917). – *Cucullaria tomentosa* G.Mey., Prim. Fl. Esseq. 13. 1818. Type: French Guiana, Richard s.n. (holotype C, isotypes G, P).
– Plate 4 E-G.

Tree 10-35 m high, trunk 40-70 cm diam. Old branchlets with cortex persistent. Leaves opposite; stipules generally lanceolate, ca. 4 mm long, pubescent; petiole 3-8 mm long, densely appressed-pubescent; blade elliptic, 6-12 x 2-3.6 cm, apex acute to acuminate, mucronate, base acute to obtuse, glabrous above, appressed-pubescent with malpighioid hairs below; primary vein impressed above, prominent below, secondary veins 9-13 per side, at 5-7 mm distance, subprominent above, slightly curved upwards, angle with primary vein 50°-60°, veinlets numerous forming a dense and conspicuous network at least below, submarginal vein absent or present only near apical third. Inflorescence terminal and axillary, up to 25 cm long; cincinni 3-4-flowered; pedicels ca. 5 mm long, densely pubescent. Spurred sepal, including calyx tube, 10-14 mm long at anthesis, spur 5-6 mm long, subconical, straight or slightly S-shaped, at an angle of ca. 30° with pedicel; petals 3, central petal provided dorsally with a narrow line of appressed hairs along central part, as long as stamen, lateral petals pubescent dorsally at least at base, ca. 1/3 shorter than stamen; staminal filament pubescent at least on ventral apical half, anther glabrous dorsally, pubescent on internal surface; staminodes nearly 0.6 x 0.1 mm, pubescent with hairs ca. 0.5 mm long; ovary glabrous, stigma lateral, ca. 1 x 1 mm. Capsule nearly 3.5 x 1 cm.

Distribution: Venezuela, the Guianas and Brazil; 52 collections studied (GU: 1; SU: 17; FG: 31).

Selected specimens: Guyana: Hancock s.n.. Suriname: Kabalebo Dam, Heyde & Lindeman 177 (U); Perica R., behind Broki, Lindeman 5047 (U); Zanderij I, BW 437 (U). French Guiana: Saül, Nouvelle France Cr., de Granville 2755 (P); Charvein, Benoist 300 (P).

Phenology: Flowering from October to December; fruiting from February to April.

Vernacular names: Suriname: alankopi (Saramacca); asjiwa, wana-kwali (Paramacca); iteballi hariraru (Arowak); wanakwari (Sranan); wosi-wosi (Carib). French Guiana: kouali, grignon, grignon fou, wana-kouali.

16. **Vochysia tucanorum** Mart., Nov. Gen. Sp. Pl. 1: 142, t. 85. 1824.

In the Guianas only: var. **tucanorum**. – *Cucullaria tucanorum* Spreng., Syst. Veg. 4. Cur. Post. 9. 1827. Type: Brazil, Minas Gerais, Martius 1179 (holotype M, isotypes BM, BR, G, K, L, M, NY, OXF, P, S, W).

Vochysia elongata Pohl, Pl. Bras. Icon. Descr. 2: 25, t. 116. 1831. – *Vochysia tucanorum* Mart. var. *elongata* (Pohl) Warm. in Mart., Fl. Bras. 13(2): 90. 1875. Type: Brazil, Minas Gerais, Pohl s.n. (BR, G,).
Vochysia tucanorum Mart. var. *microphylla* Warm. in Mart., Fl. Bras. 13(2): 90. 1875. Type: Brazil, Minas Gerais, Burchell 4571 (BR, GH, K, L, NY, OXF, P).
Vochysia opaca Pohl ex Warm. in Mart., Fl. Bras. 13(2): 91. 1875.

Tree. Leaves in whorls of 4 to many; stipules subdeltoid, up to 1.2 mm x 0.5 mm, pubescent; petiole 4-10 mm long, glabrous; blade spatulate to ovate, 3-16 x 1.5-4.5 cm, margin slightly revolute, apex obtuse-retuse to subtruncate-emarginate, base generally cuneate, glabrous on both surfaces; primary vein impressed above, prominent below, main secondary veins ca. 13 per side, angle with primary vein 45°-60°, subimpressed to subplane above, slightly prominent below, irregularly anastomosing at 1.7-2.7 mm from the margin, veinlets numerous, forming a dense network. Inflorescence terminal, up to 8 cm long; cincinni 2-4-flowered, up to 2.5 cm long; pedicels ca. 10 mm long, soon glabrescent. Spurred sepal, including calyx tube, 15-17 mm long at anthesis, spur at anthesis 7-10 mm long and nearly 1.2 mm wide at base and ca. 0.8 mm wide at subinflated apex, pendent, recurved, at an angle of 25°-30° with pedicel, in bud spur very slightly recurved, perpendicular to spurred sepal, smaller sepals ca. 3.5 mm long; petals 3, unequal, central petal ciliate at apical 1/2-2/3, nearly 1/3 longer than lateral petals and 1/4-1/3 shorter than stamen, lateral petals ciliate towards apical 1/4; staminal filament glabrous, anther ca. 1.5 x longer than filament, acute at apex and, ciliate at margin; staminodes glabrous, 0.7-10 x 0.6 mm; ovary glabrous, stigma terminal, ca. 0.5 x 0.6 mm. Capsule 2-2.4 x 0.8-1 cm.

Distribution: Brazil, Paraguay; 1 collection from Suriname: locality uncertain, Wullschlaegel 446 (BR).

Phenology: Flowering from November to July; fruiting from April to September.

Note: The other variety var. *fastigiata* Mart. has a fastigiate habit. Stafleu (1948) states that it may be a monstruosity caused by savanna fire or another agent. It may turn out to be only a form of the polymorphic *V. tucanorum*.

123 a. EUPHRONIACEAE

by

LUIS MARCANO-BERTI[2]

Trees or shrubs; indument of simple hairs. Leaves simple, alternate, petiolate; stipules small, deciduous; blades entire; penniveined. Inflorescences terminal or axillary, generally racemose; bracts deciduous. Flowers hermaphroditic, zygomorphic, perigynous; sepals 5, imbricate, subequal, inserted on a campanulate or turbinate-campanulate calyx tube or hypanthium, lacking a spur or pouch; petals 3, free, contorted, without a spur; androecium monadelphous, fertile stamens 4 in 2 opposite pairs, separated on one side by a long staminode, on the other side by 1-5 denticulate, short staminodes, anthers bithecate, opening longitudinally; ovary superior, 3-locular, ovules 2 per locule, superposed: lower one pendent, the upper one erect, style 1, simple. Capsules 3-locular, septicidal, calyx and androecium persistent; seed 1 per locule, winged at base.

Distribution: A neotropical family of a single genus with three species in Colombia, Venezuela and Brazil; one of which in Guyana.

Note: Martius described the genus *Euphronia* briefly in 1825 and included it in the family “Bonnetiae”. Shortly thereafter (1826) he gave a broader description of the genus and its type species in Nova Genera and Species Plantarum, placing it in the Spiraeaceae. In 1847 Schomburgk created *Lightia*, which was included in Trigoniaceae by Warming (1875). Hallier (1918) considered *Lightia* as a synonym of *Euphronia*. In spite of the differences between *Euphronia* or *Lightia* and the other genera that form the Trigoniaceae, it was generally accepted as part of that family, until Lleras (1976) drew attention to the illegitimate use of the name *Lightia*, this being a later synonym of *Euphronia*. Lleras re-established *Euphronia* as generic name and proposed to place it in the Vochysiaceae. Marcano-Berti (1989; 1990) placed *Euphronia* into a separate family Euphroniaceae. The family can be differentiated from Vochysiaceae and Trigoniaceae as follows:

1 Calyx with spur; corolla of 1 open or convolute petal, or 3 or 5 imbricate, free petals, sometimes absent; stamen 1, fertile and 0-2 short staminodes; ovules 2-many per locule, collateral or 2-seriate · · · · · · · · · *Vochysiaceae*
 Calyx without spur; corolla of 3 or 5, contorted free petals; stamens more than 1, and 0-6 staminodes; ovules 2 per locule, superposed or 2-many, 2-4-seriate · 2

[2] Herbarium Division, Department of Plant Ecology and Evolutionary Biology, Heidelberglaan 2, 3584 CS Utrecht, The Netherlands.

2 Corolla of 3 not spurred petals; pollen 3-colporate; disc glands absent; ovules 2 per locule, superposed: the lower one pendent, the upper one erect ··· *Euphroniaceae*

Corolla of 5 petals, the posterior one spurred; pollen 3-4-porate; disc glands present; ovules 2-many per locule, 2-4-seriate ··········· *Trigoniaceae*

LITERATURE

Lleras, E. 1976. Revision and taxonomic position of the genus Euphronia Martius ex Martius & Zuccarini (Vochysiaceae). Acta Amazonica 6: 43-47.

Marcano-Berti, L. 1989. Euphroniaceae: una nueva familia. Pittieria 18: 15-17.

Marcano-Berti, L. 1990. Descripción y tipificación de la nueva familia Euphroniaceae. Ernstia 57: 5-7.

Martius, C.F.P. von. 1825. Flora 8(1): 32.

Martius, C.F.P. von & J.C. Zuccarini. 1826. Nov. Gen. Sp. Pl. 1: 121.

Steyermark, J. 1987. A re-evaluation of the genus Euphronia (Vochysiaceae). Ann. Missouri Bot. Gard. 74: 89-94.

Schomburgk, Robert H. 1847. Beschreibung dreier neuen Pflanzen aus dem Flussgebiet des Carimani oder Camarang, eines Zuflusses des Mazaruni. Linnaea 20: 751-760.

Warming, E. 1875. Trigoniaceae. In C.F.P. von Martius, Flora Brasiliensis 13(2): 118-144.

1. **EUPHRONIA** Mart., Flora 8(1): 32. 1825; Martius & Zuccarini, Nov. Gen. Sp. Pl. 1: 121, t. 73. 1826.
Type: Euphronia hirtelloides Mart.

Lightia R.H.Schomb., Linnaea 20: 757. 1847.
Type: Lightia guianensis R.H.Schomb.

Description as for the family.

Distribution: As for the family.

1. **Euphronia guianensis** (R.H.Schomb.) Hallier f., Meded. Herb. Leiden 35: 13. 1918. – *Lightia guianensis* R.H.Schomb., Linnaea 20: 757. 1847. Type: Guyana, Carimani, Mazaruni, savanna NW of Roraima (1838 or 1839), Rob. Schomburgk I Add. ser. 2, 14 (holotype K). – Plate 6.

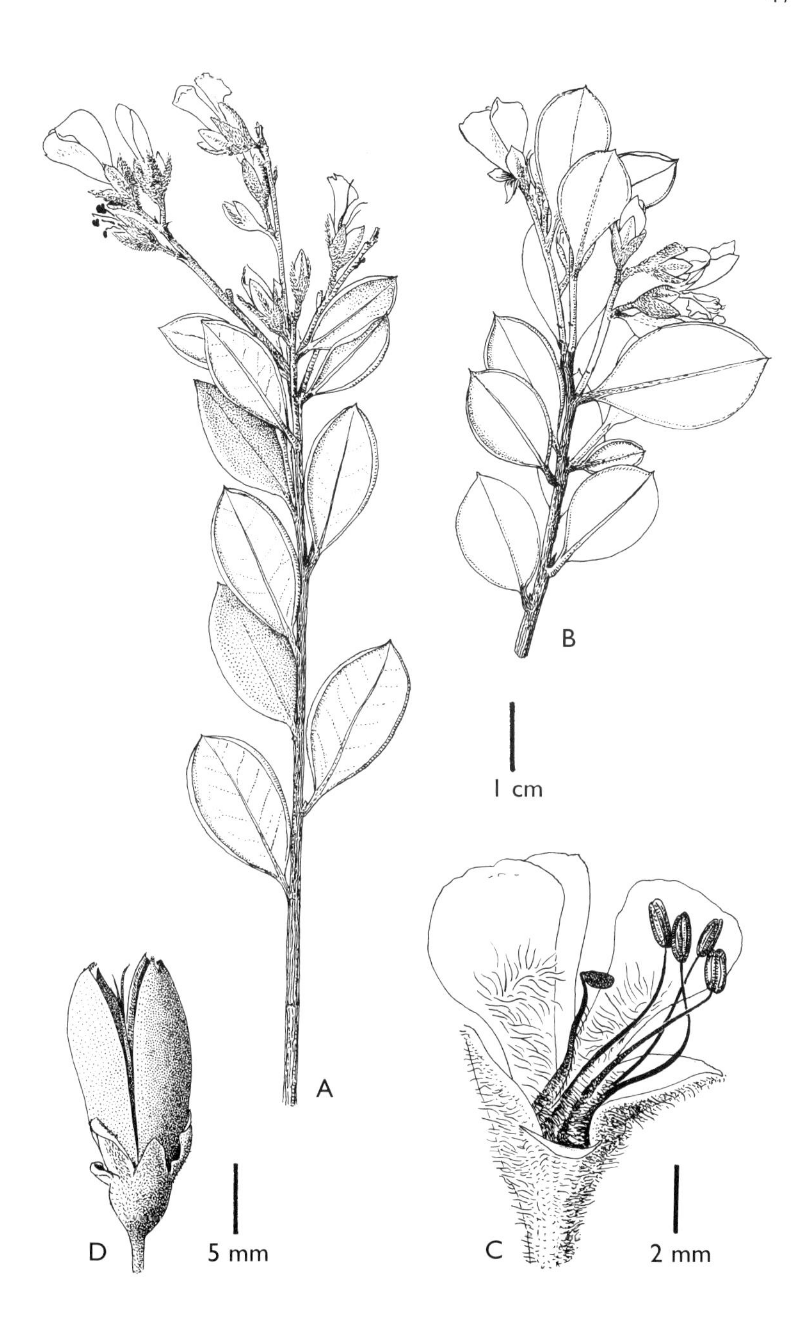

Plate 6. *Euphronia guianensis* (R.H.Schomb.) Hallier f. A, B, flowering branches; C, flower, sepal removed; D, open fruit with persistent calyx. (A, B, D, Huber 11296; C, Ule 8628). Drawing by H. Rypkema.

Shrub or tree, 0.5-15 m tall. Branchlets compressed to subcompressed, becoming terete to subterete, densely grey-arachnoid to glabrescent. Stipules deciduous, triangular to lanceolate, 1.5-4 x 0.6-1.3 mm, apex acute, base inequilateraly truncate, glabrous or glabrescent above, densely appressed-pubescent beneath; petiole 1.8-4 mm, generally densely arachnoid; blade coriaceous, obovate to elliptic, 1.2-6 x 0.65-3 cm, margin generally subrevolute to revolute, apex acute to rounded with short mucro, base cuneate to rounded, glabrous with pubescent primary vein to arachnoid above, densely greyish to whitish arachnoid beneath; primary vein flat to impressed above, prominent beneath, secondary veins 7-10 per side, 1.5-6 mm apart, lesser venation generally invisible on both sides. Raceme axillar or terminal, 1.3-6 cm long, 3-8-flowered; peduncle ca. 11 mm; bracts and bracteoles deciduous, lanceolate, 2.9-3 x 0.7-1.4 mm, apex acute, base truncate to subobtuse; pedicels 1-5 mm, pubescent (hairs to 0.8 mm resp. 0.25 mm long). Calyx, including calyx tube, 8-10 mm long, tube turbinate, 2-3 x 3-3.5 mm, densely pubescent, hairs to 0.8 mm long; outer sepals subtriangular to oblong-lanceolate, 4-5 x 2-2.5 mm, apex acute to obtuse, densely subappressed-pubescent outside, hairs to 1 mm long, sparsely pubescent inside at least towards the apical 2/3 with hairs ca. 0.25 mm long, glabrous to glabrescent towards base; inner sepals oblong to elliptic, 5.5-6.5 x 2.7-3 mm, apex subobtuse, densely appressed- or subappressed-pubescent outside along a narrow median zone, hairs 0.7-1.2 mm long, arachnoid towards margin, appressed-pubescent inside, hairs ca. 0.3 mm long; petals pink to blue, unequal, spatulate to subspatulate, 9-15 mm long, 1.5-1.7 mm wide at base, 5.2-6 mm wide near apical 1/3, apex rounded, base truncate, pubescent inside towards the basal half, hairs 1.2-1.6 mm long, appressed- or subappressed-pubescent outside towards apical half, hairs ca. 0.3 mm long; staminal tube 0.8-1 mm long, stamens in each pair different in size, filaments of larger stamens 8-10 mm, those of shorter ones 6-7.3 mm long excluding tube, retrorsely pubescent at least towards basal half, hairs 0.5-0.8 mm long, anthers oblong, glabrous, those of larger stamens ca. 2-2.5 x 0.8-1 mm, those of shorter ones 1.5-2 x 0.8-0.9 mm; long staminode: anther sterile, glabrous, ca. 1.1 x 0.35-0.5 mm, filament 5-5.9 mm long, 0.5-0.7 mm wide near base and ca. 0.08 mm near apex, retrorsely pubescent inside, at least on basal half, hairs 0.5-0.8 mm long; denticulate staminodes 2-5, subtriangular, 0.15-0.6 mm long, glabrous; ovary densely whitish woolly, ovoid to subovoid, 1.8-2.5 x ca. 1.6 mm, style 9-9.5 mm long, ca. 1 mm wide near base and ca. 0.15 mm near apex, woolly at least in basal half, less so towards apex, stigma terminal, ca. 0.6 mm wide. Fruit 1.4-2.0 x 0.6-0.8 cm, pubescent.

Distribution: Guyana and Venezuela; 11 collections studied, 1 from Guyana: Rob. Schomburgk I Add. ser. 2, 14 (K).

Phenology: Flowering from December to March; fruiting from August to December.

124. TRIGONIACEAE

by

E. LLERAS[3]

Lianas or scandent shrubs; indument of simple hairs. Branches terete, lenticellate. Leaves opposite, simple; stipules interpetiolar, often connate, deciduous or caducous; blades entire, glabrous or pubescent; pinnately veined. Inflorescences thyrses, panicles or racemes, sometimes reduced to cymules; pedicels bibracteolate. Flowers hermaphroditic, hypogynous to subperigynous, obliquely zygomorphic, plane of symmetry through third petal; receptacle of varied size and shape, slightly gibbous at base; calyx gamosepalous, base cupulate, sepals imbricate in bud, unequal; corolla papilionaceous, petals 5, contorted in bud, 2 anterior petals forming a keel, often saccate, posterior or standard petal saccate, two lateral petals or wings spatulate; androecium monadelphous, stamens 5-8 fertile, and 0-4 staminodial, unilateral, opposite the keel petals, filamental tube subperigynous, anthers basifixed, introrse, 2-locular, opening by a longitudinal slit; disc glands usually 2, opposite standard, sometimes laciniate; ovary superior, basically 3-locular, seldom 4-locular or 1-locular by reduction of parietal septa; central column absent, ovules numerous to few, style terminal, simple, stigma capitate. Capsules septicidal, valves separating from apex to base; seeds 2-several per locule, 2-4-seriate, variously pubescent.

Distribution: 31 species in four genera: one genus in Madagascar, one in Malaysia and two in the Neotropics, one of which in the Guianas.

LITERATURE

Austin, D. 1967. Trigoniaceae. Flora of Panama. Ann. Missouri Bot. Gard. 54(3): 207-210.

Grisebach, A. 1849. Linnaea 22: 27-31.

Lindeman, J.C. 1986. Trigoniaceae. In A.L. Stoffers & J.C. Lindeman, Flora of Suriname, Additions and Corrections 3(2): 550-551.

Lleras, E. 1978. Trigoniaceae. Flora Neotropica Monograph 19: 1-73.

Petersen, O.G. 1896. Trigoniaceae. In A. Engler & K. Prantl, Die Natürlichen Pflanzenfamilien 3(4): 309-319.

Reitz, R. 1967. Trigoniaceae. Flora Ilustrada Catarinense 1(13): 3-16.

[3] CPAA, Rodovia AM 010 KM 28, Manaus AM, CEP 69048 660, Caixa Postal 319, Brazil.

Stafleu, F.A. 1951. Trigoniaceae. In A. Pulle, Flora of Suriname 3(2): 174-177.
Standley, P. 1924. Trigoniaceae. North American Flora 25(4): 297-298.
Warming, J. 1875. Trigoniaceae. In C.F.P. von Martius, Flora Brasiliensis 13(2): 116-144.

1. **TRIGONIA** Aubl., Hist. Pl. Guiane 1: 387.1775.
Type: Trigonia villosa Aubl.

Hoeffnagelia Neck., Elem. Bot. 3: 68. 1790.
Nuttallia Spreng., Syst. 3: 328. 1826.
Type: Nuttallia villosa Spreng.
Mainea Vell., Fl. Flum. 275. 1829.
Type: Mainea racemosa Vell.

Description as for the family.

Distribution: 28 species, found from Oaxaca in Mexico to Paraguay, usually in lowland tropical forests. The pubescent seeds, as well as most collection localities suggest that the genus is generally restricted to forest margins or gallery forests. Commonly collected at riversides, and disturbed habitats.

KEY TO THE SPECIES

1 Secondary veins 1-6(-7) per side; veins almost glabrous ············· 2
Secondary veins 6-16 per side; veins (golden or ferruginous)strigose, tomentellous or lanate ······································· 3

2 Flowers in groups of 1-2(-3); capsule oblong, 1.5-3 cm long ··········· ···································· *4a. T. laevis* var. *laevis*
Flowers in groups of (2-)3-7; capsule obovate, 0.4-1 cm long ··········· ································ 4b. *T. laevis* var. *microcarpa*

3 Petiole 3-4 mm long ·· 4
Petiole 5-22 mm long ·· 5

4 Secondary veins 8-10 per side; sepals 2.5-3.5 mm long; standard petals 3-3.5 mm long; stamens 8 ···························· *7. T. subcymosa*
Secondary veins 5-7 per side; sepals 4-5.5 mm long; standard petal 5-7 mm long; stamens 10-12 ·················· *8a. T. villosa* var. *macrocarpa*

5 Inflorescences thyrses; stamens 5-10 · 6
Inflorescences panicles or racemes; stamens 10-12 · · · · · · · · · · · · · · · · · · 8

6 Tertiary veins evident; bracts, bracteoles, and sepals papillose-glandular; disk glands not lobed; leaf margin slightly revolute · · · · · · *6. T. reticulata*
Tertiary veins not especially evident; bracts, bracteoles, and sepals not papiiiose-glandeular; disk glands bilobed; leaf margin plane · · · · · · · · · 7

7 Leaf golden-strigose beneath; branches ferruginous-strigose · *2. T. coppenamensis*
Leaf tomentellous on the veins beneath, glabrescent; branches tomentellous, glabrescent · *1. T. candelabra*

8 Leaves pubescent above · · · · · · · · · · · · · · · · · · · *8b. T. villosa* var. *villosa*
Leaves glabrous above · 9

9 Secondary veins 7-8 per side; branches strigose, glabrescent; fruit ribbed, glabrous · *3. T. hypoleuca*
Secondary veins 10-16 per side; branches lanate, glabrescent; fruit velutinous-tomentose to slightly strigose · · · · · · · · · *5. T. nivea* var. *nivea*

1. **Trigonia candelabra** Lleras, Fl. Neotropica 19: 41, Fig. 14, 15. 1978. Type: Suriname, Brokopondo, Brokopondo village, van Donselaar 2812, fl. (holotype U). – Plate 7.

Liana, young branches tomentellous, glabrescent. Stipules caducous, not seen; petiole 8-15 mm long, tomentellous; blade subcoriaceous, elliptic to obovate, 8-21 x 4-10 cm, margin entire, plane, apex acute to acuminate, base obtuse to oblique, glabrous above, slightly tomentellous beneath, glabrescent; primary vein plane above, prominent beneath, strigulose, secondary veins 6-9 pairs. Inflorescences terminal and subterminal axillary thyrses, 5-9 cm long; flowers in compound dichasia, axes 1-7 mm long, tomentellous; peduncles 0.8-2 mm long, bracts subulate, 1.5-2.5 mm long, tomentellous; pedicels 1-2.5 mm long; bracteoles subulate to ovate, 0.8-1.5 long. Sepals ovate to oblong, 4.3-4.6 x 1-2 mm, tomentellous; petals white, standard 4-5 x 2-3 mm, saccate to the middle, apex slightly revolute, throat barbate, wings spatulate, 4-4.5 x 1.3-1.6 mm, glabrous, keel petals saccate, 3-3.5 x 1-1.3 mm; stamens 8-10, 2-4 staminodial, filaments connate to the middle, 1.5-1.8 mm long, anthers oblong to ovate, 0.8 x 0.6-0.9 mm, apex acuminate; disc glands 2, 2-3-lobed, rectangular, ca. 0.5 mm wide, laciniate; ovary subglobose, ca. 1 mm in diam., 3-locular, villous, ovules numerous, style erect, 1.3-1.6 mm long, glabrous to villous, stigma round, ca. 0.3 mm in diam. Capsule oblongoid-obovoid, 6.5-8 cm long, ca. 1 cm wide, externally velutinous-tomentose, glabrous internally; seeds ca. 10 per locule, subglobose, ca. 4 mm in diam., trichomes ca. 7 mm long.

Plate 7. *Trigonia candelabra* Lleras. A, habit vegetative branch; B, habit, flowering branch; C, habit, fruiting branch; D, flower; E, ovary; F, keel petal; G, wing; H, standard; I, anthers (twice the magnification of other flower parts); J, compound dichasium in bud.

Distribution: Known from 2 collections in Suriname and 2 in Pará, Brazil, from forest edges or secondary growth.

Specimens examined: Suriname: Brokopondo, the type specimen and van Donselaar 2891 (U).

2. **Trigonia coppenamensis** Stafleu, Rec. Trav. Bot. Neerl. 42: 70. 1950. Type: Suriname, Schmidt Mt., headwaters of the Coppename R., Maguire 24857 (holotype U, isotypes F, NY).

Liana, branches striate (almost angular), not conspicuously lenticellate, ferruginous-strigose. Stipules triangular, 4-7 x 2-3 mm, strigose; petiole 10-22 mm long, striate to terete, strigose; blade subcoriaceous, elliptic to obovate, sometimes almost circular, 5.5-15 x 2.5-12 cm, margin entire, apex acute to acuminate, base subrotund; venation eucamptodromous, densely golden-strigose (including tertiary veins beneath), primary vein prominulous above, prominent beneath, secondary veins 6-9 pairs, plane above, prominulous beneath, tertiary veins impressed above, prominulous beneath. Inflorescences terminal and subterminal, axillary thyrses, to 12 cm long; flowers in dichasia of 4-6; peduncles 4-8 mm long, ca. 0.5 mm thick, strigose; pedicels 2.5-5.5 mm long, ca. 0.5 mm thick, strigose; bracts and bracteoles 3-6 x 1-2 mm, linear-ovate (almost navicular), revolute at margin, curved upwards towards apex of flower, strigose. Flower buds to 6 mm long. Sepals ovate to oblong, ca. 5 mm long, thickened at veins (appearing striate), strigose; standard with pouch extending to 1/2 the length, throat barbate, wings spatulate, barbate at base, keel petals saccate; fertile stamens 6, staminodes 3-4, anthers with short knob-like projection at apex; disc glands 2, 2-3-lobed; ovary subglobose, densely villous, ovules numerous, style glabrous, stigma 3-lobed. Fruit unknown.

Distribution: Only known from the type collection from Suriname.

3. **Trigonia hypoleuca** Griseb., Linnaea 22: 30. 1849; Warm. in Mart., Fl. Bras. 13(2): 140. 1875. Guyana. Type: Rich. Schomburgk 315, fl. (holotype GOET, isotype K).

Trigonia hypoleuca var. *pubescens* Warm. in Mart., Fl. Bras. 13(2): 140. 1875. Type: Suriname, Wullschlaegel 8161, fl., fr. (holotype BR, isotypes IAN, NY, VEN).
Trigonia xanthopila Garcke, Linnaea 22: 51. 1849. Syntype: Suriname, bank upper Saramacca R., Cassipoera Cr., Kegel 1177, fl. (holotype on 2 sheets as indicated by Garcke, GOET).

Liana or scandent shrub, branches terete, densely lenticellate, young ones strigose, becoming glabrous with age. Stipules subulate, 2-3 mm long, caducous; petiole 5-12 mm long, rugulose, almost glabrous; blade subcoriaceous, oblong-elliptic or obovate, sometimes unequilateral, 8-18(-20) x 4-10 mm, margin entire, apex acute to acuminate, base cuneate to obtuse, intercostal pubescence absent above, lanate beneath; venation eucamptodromous, strigose-pubescent, primary vein plane above, prominulous beneath, secondary veins 7-8 pairs. Inflorescences terminal and subterminal axillary panicles, to 30 cm long; flowers in groups of 1-4; peduncles and pedicels 2-4.5 mm long, strigose; bracts and bracteoles linear, 1-2.5 mm long, densely strigose. Sepals ovate to oblong, 4.8-6 x 1.5-3.0 mm, strigose where exposed, lanate where protected; standard 5.5-8.5 x 4-5 mm, pouch extending along to the middle, upper portion revolute, throat barbate, wings broadly spatulate, 5.5-8.5 x 2.8-5 mm, barbate at base, keel petals 5-7 x 2.2-3.5 mm, pouch extending to the middle, barbate at base; stamens 10-11, 3-4 staminodial with terminal globose knobs, filaments 1.5-1.8 mm long, free to the middle, anthers obovate, 0.5-0.7 x 0.3-0.4 mm; disc glands 2, 2-3-lobed, lobes 2-3 mm high, upper half formed of subulate, pointed, strigose laciniae; ovary subglobose, 1.0-1.5 mm in diam., barbate-pilose, 1-locular, ovules numerous, style clavate, 1.5-1.8 mm long, glabrous, stigma ca. 0.3 mm in diam. Capsule 5-7 x 2.5-3 cm, exocarp coriaceous with prominulous veins, glabrous, endocarp ending flush with the exocarp, cartilaginous; seeds elliptic, 6-8 mm long, flattened, equinate-pubescent.

Distribution: Found mainly along forested, periodically flooded river banks in the Guianas and N Brazil; 26 collections studied (GU: 9; SU: 15; FG: 2).

Selected specimens: Guyana: Gleason et al. 549 (GH, NY, US); Jenman 1296 (K). Suriname: Brokopondo Distr., Suriname R., S of Gansee, van Donselaar 1335 (U); Nickerie Distr., Kabalebo Dam area, Heyde & Lindeman 123 (U); French Guiana: Béna 4204 (U); Ile Portal, Sagot s.n. (F, G, GH, NY, US).

4. **Trigonia laevis** Aubl., Hist. Pl. Guiane 1: 390; 3: pl. 150. 1775. Type: Aublet s.n., French Guiana, Cayenne (holotype BM, isotypes F, P).

Liana, branches terete, lenticellate, young ones strigulose, glabrescent. Stipules subulate, 1.5-2 mm long, strigulose; petiole 4-8 mm long, terete to canaliculate, strigose or glabrous; blade chartaceous to subcoriaceous, broadly elliptic to oblong-elliptic, sometimes unequilateral, 4-9 x 2.2-4.5

cm, margin entire, apex acute to acuminate, base oblique to obtuse, intercostal pubescence absent; venation eucamptodromous, glabrous or slightly pilose, primary vein prominulous above, prominent beneath, secondary veins 4-6(-7) pairs, prominulous on both surfaces. Inflorescences terminal and subterminal panicles (sometimes thyrses) to 30 cm long; flowers in groups (cincinni) of 1-7; peduncles 0.2-0.5 mm long, strigose, bracts subulate to linear, 1-3 mm long, strigose; pedicels 0.8-2 mm long, strigose; bracteoles subulate to linear, 0.3-1 mm long, strigose. Sepals deltoid to oblong, 2-3 x 0.8-1.5 mm, strigose; standard 2.5-5 x 2-3 mm, the pouch extending along 2/3 of the length, throat barbate, wings spatulate, 2.5-5 x 0.5-1 mm, barbate at base, keel petals 2-3.5 x 1.5-2 mm, sometimes lanate at apex; fertile stamens 5-7, staminodes 1-3, filaments 0.7-1.2 mm long, connate for 2/3 of the length, anthers subglobose, 0.4-0.5 mm long, ca. 0.4 mm wide, staminodes sometimes with small, terminal laciniate appendage; disc glands 2, 2-3-lobed, irregular, ca. 0.3 mm, glabrous; ovary subglobose, 0.6-0.8 mm in diam., villous, ovules numerous, style slightly villous, 1-1.3 mm long, stigma 3-lobed, ca. 0.2 mm in diam. Capsule oblong or obovate, 0.6-3 x 0.6-1 cm, exocarp thin, yellow velutinous when young, glabrescent, endocarp woody, glabrous; seeds 1-4 per capsule, subglobose, ca. 0.3 mm in diam., barbate-villous, trichomes ca. 10 mm long.

Note: The species is separated into two varieties, both of which occur in the Guianas.

4a. **Trigonia laevis** Aubl. var. **laevis**

> *Trigonia kaieteurensis* Maguire, Bull. Torrey Bot. Club 75: 399. 1948. Type: Guyana, Kaieteur Plateau, Maguire & Fanshawe 23192, fr. (holotype NY, isotypes A, BR, F, MO, NY, P, US, VEN).

Flowers in groups of 1-3; petals 2.5-3.5 mm long. Capsule oblong, 1.5-3 cm long, ca. 1 cm wide; seeds usually 3-4 per locule.

Distribution: Known only from few collections in Guyana and French Guiana; 9 collections studied (GU: 4; FG: 5).

Selected specimens: Guyana: Mabura Hill, Maas et al. 7124 (U); Maguire et al. 23192. French Guiana: Piste St. Elie, Prévost 1165 (CAY); St. Laurent region, Skog & Feuillet 7445 (US).

4b. **Trigonia laevis** Aubl. var. **microcarpa** (Sagot ex Warm.) Sagot, Ann. Sci Nat. Paris Ser. 6.11: 176. 1881. – *Trigonia microcarpa* Sagot ex Warm. in Mart., Fl. Bras. 13(2): 131. 1875. Type: French Guiana, Karouany, Sagot 36, fl., fr. (holotype P, isotypes BR, F, GH, GOET, P, W).

Trigonia parviflora Benth., J. Bot. (Hooker) 3: 163. 1851. Type: Brazil, Pará, Santarem, Spruce s.n., fl. (holotype K, isotypes C, CGE, F, G, GOET, M, NY, P, W), nomen illegit.

Flowers in groups of 3-6; petals usually 3.5-5 mm long. Capsule obovate, 0.6-1 cm long and wide; seeds usually 1-2(-3) per locule.

Distribution: A widespread rainforest variety, found from French Guiana to Bolivia; 30 collections studied (GU: 6; SU: 15; FG: 9).

Selected specimens: Guyana: Mazaruni Station, Sandwith 563 (G, NY, P, U, US); A.C. Smith 2752 (A, G, MO, NY, U, US). Suriname: Kabalebo Dam area, Lindeman & Görts-van Rijn et al. 58 (U); Maguire 24858. French Guiana: Perrottet 262, Mélinon 230 (P).

5. **Trigonia nivea** Cambess. in St. Hil., Fl. Bras. Mer. 2: 81. 1829. Type: Brazil, St. Hilaire 226, fl., fr. (holotype MPU).

In the Guianas only: var. **nivea**

Liana, branches terete, lenticellate, lanate when young, glabrescent. Stipules triangular, 6-10 mm long, strigulose, caducous; petiole 2-8 mm long, strigose or lanate, sometimes becoming glabrous; blade subcoriaceous, elliptic to oblong-elliptic, sometimes ovate or obovate, 5-13 x 1-4.5 cm, margin entire or revolute, apex acute to acuminate, sometimes mucronate, base obtuse, intercostal pubescence absent above, lanate to appressed-lanate (almost sericeous) beneath; venation eucamptodromous, slightly strigulose or glabrous above, lanate or strigose beneath, primary vein plane above, prominent beneath, secondary veins 10-16 pairs. Inflorescences terminal and axillary racemes or panicles, 5-10 cm long, varying from highly congested to very open (sometimes on the same plant); flowers in groups (cincinni) of 1-4; peduncles 0-5 mm long, ca. 0.5 mm thick, strigose; bracts subulate or triangular, 2-6 mm long, strigose; pedicels 1-5 mm long, strigose; bracteoles subulste, 1-4 mm long, strigose. Sepals ovate or oblong, 3-5 x 1.5-3 mm, lanate where protected, strigose where exposed; standard 4.5-6.5 x 3 mm, pouch extending to the middle, throat barbate, keel petals 3.5-5 x 3 mm, pouch extending along 2/3 of the length; stamens

10-11, 3-4 staminodial, filaments 1-2 mm long, free to the middle, anthers oblong, 0.4-0.7 x 0.3 mm; disc glands 2-3-lobed, trapezoid or rounded, ca. 0.4 mm wide, sometimes with short laciniae, glabrous; ovary subglobose, 1-1.5 mm in diam., densely villous, ovules numerous, style 1-2 mm long, glabrous, stigma 3-lobed, 0.2-0.3 mm in diam. Capsule 3-7 x 1-1.5 cm, exocarp thin, velutinous-tomentose or slightly strigose, endocarp separable from mesocarp, velutinous-tomentose or partially so, sometimes glabrous; seeds ovate, barbate-pubescent, trichomes ca. 10 mm long.

Distribution: Variety *nivea* grows in coastal and inland forests and grasslands from Venezuela to S Brazil and Paraguay. The two other varieties are restricted to parts of Brazil; 3 collections studied from Guyana.

Specimens examined: Guyana: Rupununi, Irwin 679 (US); Rupununi R., Monkey Pond landing, Maas et al. 7392 (U); McConnell & Quelch 222 (K).

6. **Trigonia reticulata** Lleras, Fl. Neotropica 19: 41. 1978. Type: Guyana, upper Mazaruni R. Basin, Kako R., Tillett & Tillett 45524, fl. (holotype NY, isotypes F, GH, P, US).

Liana, branches lenticellate, tomentellous when young, glabrescent. Stipules caducous, not seen; petiole 9-18 mm long, tomentellous; blade subcoriaceous, ovate to elliptic, sometimes obovate, 4-11 x 3-6.5 cm, margin slightly revolute, apex acute or acuminate, base oblique to subrotund, glabrous above, lanate-tomentellous beneath; primary vein plane above, prominent beneath, strigulose, secondary veins (8-)9-11(-12) pairs, tertiary and lesser veins reticulate, evident beneath. Inflorescences terminal and axillary thyrses, 4-15 cm long; flowers in simple or compound dichasia, their axes 0.5-2 mm long, tomentellous; bracts subulate, 1-2.5 mm long, margin glandular-papillate; pedicels 1-2.5 mm long, tomentellous; bracteoles subulate, 1-1.5 mm long, margin as of bracts. Sepals ovate to oblong, 2.8-4 x 1.7-2.2 mm, sometimes glandular, tomentellous; standard 4.3-4.6 x 2.5-2.8 mm, saccate to the middle, apex slightly revolute, throat barbate, wings spatulate, 3.3-3.8 x 1.3-1.6 mm, barbate at base, keel petals saccate, 2.8-3 mm x 1.-1.8 mm; stamens 8-10, 2-4 staminodial and appendiculate, filaments connate for 1/2 their length, 1-1.4 mm long, anthers oblong to elliptic, 0.5-0.7 x 0.3-0.4 mm; disc glands 2, not lobed, labiate, lanate-villous; ovary subglobose to pyramidal, ca. 0.6 mm thick, 3-locular, villous, ovules numerous, style erect, 1-1.2 mm long, stigma rounded, ca. 0.3 mm in diam., white. Young fruit oblong, densely villous-tomentose.

Distribution: Known from a few collections made in open areas along roads and rivers in Guyana and Venezuela; 2 collections studied from Guyana.

Specimens examined: Guyana: the type collection and Kukui R., FD 7992 (NY).

7. **Trigonia subcymosa** Benth., London J. Bot. 2: 373. 1843. Type: Guyana, Essequibo, Rob. Schomburgk I 56 earlier sets, fl. (lectotype K, isolectotypes CGE, G, NY, W, paralectotype Rob. Schomburgk I later sets).

Large shrub, branches terete, densely lenticellate, strigose when young, glabrescent. Stipules lunulate, 4-5 x 2 mm, caducous; petiole 3-4 mm long, densely golden-brown strigose; blade chartaceous, elliptic to obovate or oblong, 3-8 x 2-4 cm, margin entire, apex obtuse, acute or mucronate, base cuneate to obtuse, intercostal pubescence very sparsely strigose above, depressed-lanate beneath, mostly cream or yellowish; venation eucamptodromous, sparsely golden strigose, primary vein plane above, prominulous beneath, secondary veins 8-10 pairs. Inflorescences terminal, pyramidal panicles, to 15 cm long; flowers in cymules of 2-4; peduncles 1-8 mm long, diminishing in length towards apex of inflorescence; pedicels 1-2 mm long, villous-strigose; bracts and bracteoles, 2-3 x 0.5-0.7 mm, ovate, curved upwards, villous-strigose. Sepals ovate or oblong, 2.5-3.5 x 0.7-1.5 mm, strigose or villous-strigose where exposed, lanate where protected; standard 3-3.5 x 2 cm, pouch extending to the middle, erect along upper portion or nearly so, apex revolute, throat barbate, wings spatulate, 2.5-3.1 x 0.8-1 mm, glabrous at base, keel petals 2.5-2.8 x 2 mm, pouch extending from 1/3 of the length up, apex revolute, glabrous; stamens 8, 2-3 staminodial, filaments 1-1.5 mm long, free for 1/3 of the length, anthers ovate or oblong, 0.4-0.5 x 0.2 mm; disc glands 2, 2-3-lobed, deltoid or trapezoid, ca. 0.4 mm wide, glabrous; ovary subglobose, ca. 0.4 mm in diam., villous-barbate, ovules numerous, style 0.8-1.1 mm long, glabrous or villous, stigma 3-lobed, ca. 0.1 mm in diam. Fruit unknown.

Distribution: Known only from 3 Schomburgk collections made in Guyana: the syntypes from the Essequibo and Rob. II 249 or Rich. 373 are without locality.

8. **Trigonia villosa** Aubl., Hist. Pl. Guiane 1: 338; 3: pl. 149. 1775. Type: French Guiana, Cayenne, Aublet s.n. “Cayenne, 1775”, fl, fr (lectotype BM).

Liana, branches terete, slightly lenticellate, densely golden-brown strigose when young, glabrescent. Stipules coriaceous, subulate, 3-6 x 1.5-2 mm, caducous; petiole 3-10 mm long, very densely golden-brown strigose; blade subcoriaceous, broadly elliptic to obovate, sometimes oblong, inequilateral, 5-14 x 2-8.5 cm, margin entire, apex acute to acuminate, base cuneate to obtuse, intercostal pubescence very sparsely strigose above, yellowish or greenish-white lanate beneath; venation eucamptodromous, densely golden-strigose pubescent, primary vein plane above, prominulous to prominent beneath, secondary veins 6-9 pairs, inserted at angles of ca. 70° to primary vein. Inflorescences terminal and subterminal, axillary panicles, to 25 cm long; flowers in groups (cincinni) of 1-2(-3); peduncles 0.5-3 mm long, strigose; bracts subulate to linear, sometimes ovate, 2-3 x 0.5-1.5 mm, strigose; pedicels 1.8-2.8 mm long, strigose. Sepals ovate or oblong, 4-5.5 x 2-3.5 mm, acute or rounded at apex, lanate where protected, strigose where exposed; standard 5-7 mm long, pouch extending to the middle, upper half revolute, irregular at apex, throat barbate, wings spatulate, 6-6.5 x 2-3 mm, barbate at base, keel petals 4-5 x 3-4 mm, the pouch extending along 2/3 of the length, barbate at base; stamens 10-11(-12), 3-4 staminodial, filaments 2.5-3 mm long, free for 1/3 of the length, anthers obovate or oblong, 0.5-0.7 mm long, ca. 0.4 mm wide; disc glands 2, 2-3-lobed, lobes deltoid, 0.2-0.3 mm wide, glabrous; ovary subglobose, ca. 1 mm in diam., barbate-pubescent, ovules numerous, style 2.5-2.8 mm long, glabrous or slightly villous, stigma 3-lobed, ca. 0.2 mm in diam. Capsule 4.5-11 cm long, exocarp thin, sometimes brown velutinous-tomentose pubescent, to glabrous or nearly so; seeds ca. 20 per locule, ovoid, barbate-pubescent, trichomes to 20 mm long, spirally disposed.

Note: The species is divided into two varieties, both of which occur in the Guianas.

8a. **Trigonia villosa** var. **macrocarpa** (Benth.) Lleras, Fl. Neotropica 19: 57. 1978. – *T. macrocarpa* Benth., London J. Bot. 2: 373. 1843. Type: Guyana, Essequibo R., Rob. Schomburgk 54, fl., fr. (holotype K, isotypes C, CGE, F, G, NY, US, W).

Trigonia macrostachya Klotzsch in M.R. Schomburgk, Reisen Brit.-Guiana 3: 1183. 1849 (nomen nudum); Grisebach, Linnaea 22: 28, 1849 for Rich. Schomburgk 343, Essequibo R.

Leaves with petiole 3-4 mm long. Capsule over 9 cm long, exocarp usually dirty yellow-green, endocarp glabrous or nearly so.

Distribution: This variety is known from a few collections only in riverine forests in Guyana and Brazil (Amazonas and Ceará); the 2 type collections studied from Guyana.

Specimens examined: Rob. Schomburgk ser. I 54 earlier sets; Rich. Schomburgk 343.

8b. **Trigonia villosa** Aubl. var. **villosa**

Trigonia villosa Aubl. var. *angustifolia* Benth., London J. Bot. 2: 373. 1843. Syntypes: Guyana, Essequibo, Rob. Schomburgk I 63, earlier sets and 54 later sets.
Trigonia villosa Aubl. var. *obtusata* DC., Prodr. 1: 571. 1824. Type: French Guiana, Perrottet 261, fl. (holotype G, isotype G).
Trigonia villosa Aubl. var. *cuneata* DC., Prodr. 1: 571. 1824. Type: French Guiana, Perrottet 259, fl. (holotype G, isotype G).
Trigonia villosa Aubl. var. *oblonga* DC., Prodr. 1: 571. 1824. Type: French Guiana, Perrottet 260, fl. (holotype G, isotype G).
Trigonia mollis DC., Prodr. 1: 571. 1824. Type: Brazil, Rio de Janeiro, Corcovado, Martius 179, fl. (holotype G, isotype M).
Trigonia cepo Cambess. in St. Hil., Fl. Bras. Merid. 2: 115. 1829. Type: Brazil, Rio de Janeiro, St. Hilaire 125, fl. (holotype MPU, isotypes F, G, P, US).

Leaves with petiole 5-10 mm long. Capsule 4.5-7.5 cm long, exocarp usually reddish-brown, endocarp densely velutinous- tomentose.

Distribution: A widespread variety, known from the Guianas, Amapá and Pará in northern Brazil, and as a significant range disjunction, from Rio de Janeiro; 28 collections studied (GU: 5; SU: 3; FG: 20).

Selected specimens: Guyana: Essequibo R., Jenman 1083 (U), 6736 (NY, U). Suriname: Corantijn R., near Wonotobo, BW 2864 (U); Kabalebo Dam area, Lindeman & Görts-van Rijn et al. 98, 355 (U). French Guiana: Ile de Cayenne, de Granville BC-69 (CAY); F. Hallé 847.

DOUBTFUL SPECIES

Trigonia spruceana Benth. ex Warming

A collection Pipoly 8656, that came after the treatment had been completed, may be referable to *T. spruceana*. There are no further records for the Guianas. The species is known from the basin of the Rio Negro, occurring in Venezuela and Brazil; in periodically flooded riverside forest especially along black water rivers.

126. KRAMERIACEAE

by

BERYL B. SIMPSON[4]

Perennial shrubs or subshrubs. Leaves estipulate, alternate, simple, entire, linear to ovate, acute, mucronate, decreasing in size apically. Inflorescences of terminal or axillary racemes; flowering stalks of peduncles and pedicels demarcated by a pair of opposite bracteoles with pedicel only deciduous. Flowers zygomorphic; sepals showy, 4 (5 in other species), entire, free, with the lowermost sepal cupped, uniformly pink, purple, rose, or bicolored with the upper sepals pink and the lower sepals fading to white; petals 5(4), dimorphic, lower 2 glandular petals flattened, fleshy, orbicular to cuneate, ca. 2-4 mm long, flanking the ovary, dorsally covered with saccate blisters containing secreted oils, remaining petaloid petals clawed, inserted at base of upper side of ovary, free or connate basally; stamens 4, inserted at, or on, bases of petaloid petals, free beyond point of insertion, didynamous; ovary superior, ovoid, densely pubescent, appearing monocarpellate due to ontogenetic suppression of one of the two carpels. Fruits globose, variably vestitured, spinose, spines with basal trichomes and retrorse barbs along shaft; seed 1, lacking endosperm, globose, seed coat gray-brown, smooth.

Distribution: A monogeneric family with 18 species distributed in North, Central, and South America and the West Indies with the greatest concentrations of species in the arid and semi-arid regions of N Mexico and in central and E Brazil. Most species are found at low elevations, but one occurs at altitudes over 3500 m in Peru, Bolivia, and N Argentina. All species occur in open arid or seasonally dry habitats primarily on rocky or sandy soils. In the Guianas: two species, a third species to which several Guyana specimens have been erroneously referred might eventually be found in the Guiana region and is included here.

LITERATURE

Simpson, B.B. 1989. Krameriaceae. Fl. Neotrop. Monogr. 49: 1-108.
Simpson, B.B. 1991. The past and present uses of rhatany (Krameria, Krameriaceae). Econ. Bot. 45: 397-409.

[4] Department of Botany, The University of Texas, Austin, TX 78713, U.S.A.

1. **KRAMERIA** Loefl., Iter Hispan. 195. 1758.
   Type: Krameria ixine Loefl.

Characters as for the family.

KEY TO THE SPECIES

1 Openly branched, erect shrubs, 0.6-2.0 m tall; inflorescences compound, racemose; leaves of major branches more than 3 mm wide; petiole more than 2 mm long; sepals pink, often turning white with age giving the flowers a pink and white appearance; fruit body sericeous or tomentose, with red, orange, or yellow-tipped spines of 0.25 mm or less in diam. at base and bearing retorse barbs along the shaft · 2

Lax, sprawling or decumbent perennials; inflorescences secund; leaves of major branches less than 3 mm wide, attenuate at the base; petiole 1 mm long; sepals purple, pink, or rose, not fading to white; fruit body lanose, with tan spines of 0.5 mm in diam. at base and bearing a few retorse teeth apically · *2. K. spartioides*

2 Leaves of major stems lanceolate, 3-10 mm wide, terminated by an apicule of 1-2 mm long, shortly villose; stem tips densely strigose; racemes with lateral branches obliquely branching from the main axis; flowering stalks subtended by bracts over 5 mm long or longer · · · · · · · · · · · · *1. K. ixine*

Leaves of major stems broadly lanceolate to ovate, 5-15 mm wide, terminated by an apicule less than 1 mm long, densely tomentose-lanose; stem tips completely covered with golden brown lanose trichomes; racemes robust with lateral branches divergent at right angles from the main axis; flowering stalks subtended by bracts less than 5 mm long; species not yet reported from the Guianas · *3. K. tomentosa*

1. **Krameria ixine** Loefl., Iter Hispan. 195. 1758. Type: Venezuela, Sucre, 18 km SE of Cumaná on the rd. to Cumanocoa, Simpson 8504 (neotype BM; isotypes NY, TEX, US). – *Krameria ixina* L., Sp. Pl. ed. 2: 177. 1762, an orthographic variant of Loefling's name, K. ixine.

Upright, openly-branched shrubs 0.3-1.5 m tall; young stems green, densely strigose at the tips. Leaves of major branches lanceolate to elliptic, 17-36 x 3-8 mm, including a 2-9 mm long petiole, mucronate with the apicule 1-2 mm long, both surfaces shortly villous with primary vein rarely visible below. Racemes lateral and terminal with lateral branches oblique to main axis; flowering stalks subtended by bracts generally longer than 5 mm, separated medially into peduncle and pedicel by a pair of small linear, 3-5 mm long bracteoles; sepals pink or rose fading to pale pink or white with age, densely sericeous dorsally; glandular petals pink or rose-colored, petaloid petals pink, 4.5-7.0 mm long, connate basally for 2 mm; stamens 4 (rarely 3), inserted on the connate claws of the petaloid petals, longer pair 3.0-5.0 mm long. Fruit body green, 4-7 mm in diam. excluding spines, variably strigose, spines red or orange, 2.8-4.6 mm long, 0.25 mm or less in diam. at base, with retrorse barbs along shaft, but most pronounced distally, sometimes with a few long trichomes basally.

Distribution: Scattered in open grass or oak or pine savannas along the west coast of Mexico from Sinaloa to Oaxaca, in E Guatemala, S Honduras, and in Guanacaste Province of Costa Rica. In the Caribbean, the species has been reported on Hispaniola and south through most of the Greater and Lesser Antilles, and in South America in dry, grassy areas in NE Colombia, N Venezuela and the N Rupununi Savanna in Guyana. Most populations occur near sea level, but in Guatemala the species reaches elevations of 1500 m. (GU: 7).

Specimens examined: Guyana, Rupununi Distr.: N of Rupununi, Davis 873 (K, NY); Manari, Maas et al. 3655 (MO, NY, U); S of Lethem, Irwin 581 (US) and 885 (TEX); Lethem, Jansen-Jacobs et al. 578 (U); St. Ignatius, Harrison 1320 (K).

Phenology: Recorded flowering and fruiting times in the flora area are April, July and October.

Uses: No uses have been reported in Guyana, but the species has been extensively used medicinally elsewhere (see Simpson, 1991).

2. **Krameria spartioides** Klotzsch ex O. Berg, Bot. Zeitung (Berlin) 14: 761. 1856. Type: Colombia, Coyaima, along the Río Magdalena, Goudot s.n. (lectotype W, isolectotypes K, P). – Plate 8.

Sprawling subshrubs to 0.6 m tall; stems strigose at the tips. Leaves of major stems linear to linear-lanceolate, attenuate, 21-30 x 1.5-2.5 mm, including a 1 mm long petiole, mucronate with the apicule to 1 mm long, variably strigose on both surfaces; primary vein apparent. Racemes terminal, secund; flowering stalks subtended by bracts shorter than 5 mm, separated in the proximal half into peduncle and pedicel by a pair of linear, 4.0-5.5 mm long bracteoles; sepals pink-purple, sparsely sericeous dorsally; glandular petal color unreported, petaloid petals with color unreported, 6.5-8.5 mm long, connate basally for 3.4-4.5 mm; stamens 4, inserted on the connate claws of the petaloid petals, longer pair ca. 5 mm long. Fruit body green-maroon, 4-6 mm in diam. excluding spines, lanose, spines tan, stout, 0.5-2.0 mm long, 0.5 mm in diam. at base, minutely barbed apically and sparsely strigose basally.

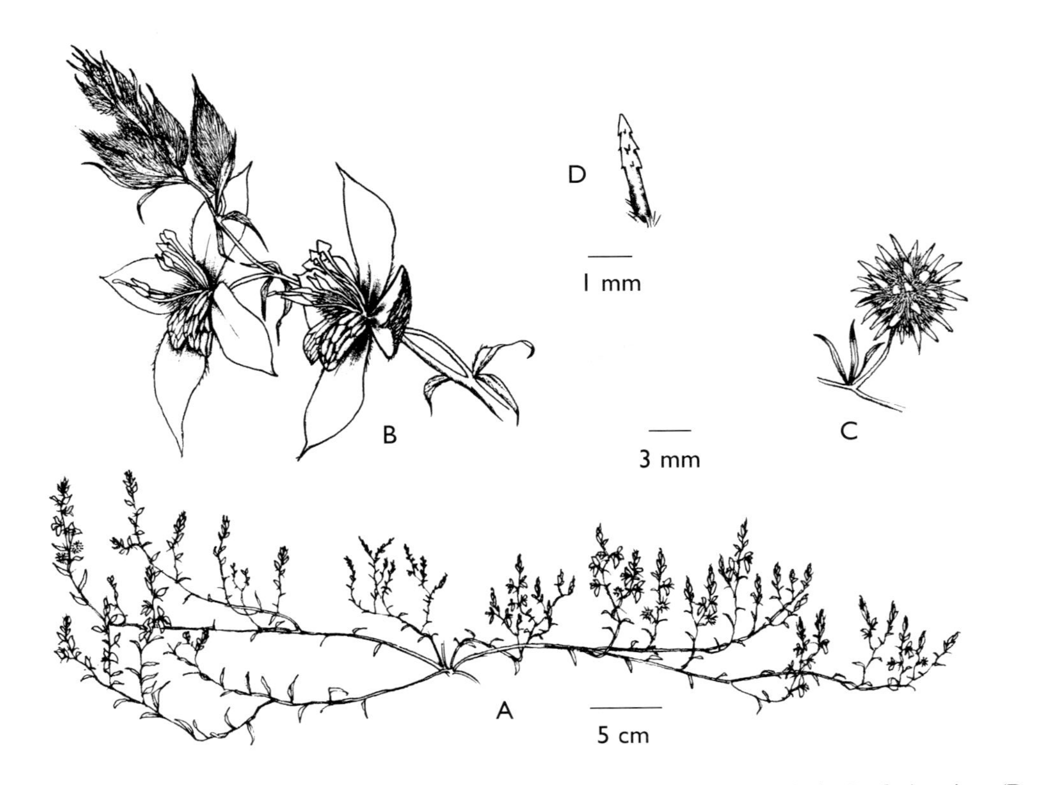

Plate 8. *Krameria spartioides* Klotzsch ex O. Berg. A, habit; B, tip of a flowering shoot; C, fruit; D, fruit spine. (Drawn from Rob. Schomburgk II 496 and Jansen-Jacobs et al. 473).

Distribution: Apparently disjunct in savannas between extreme N and N-C Colombia and the boundary region between Venezuela, Brazil and Guyana (GU: 7).

Specimens examined: Guyana: Roraima, Rob. Schomburgk I 201S (K) and II 496 (BM, G, K, P), Rich. Schomburgk 570 (G); S Pakaraima Mts., along Ireng R. mouth and Corona Falls, savanna ca. 9 km NW of Karasabai, Hoffman 1063 (US); Essequibo R., S Rupununi, Davis 1701 (K, NY); Rupununi Savanna, Mountain point near the ranch of Shirley Humphries, Jansen-Jacobs et al. 473 (MO, U); without locality, Wilson-Browne 30 (U).

Phenology: Flowering and fruiting specimens collected from the flora area in February and October.

3. **Krameria tomentosa** A.St.-Hil., Ann. Soc. Sci. Orléans, Mém. sér. 2. 9: 18. 1828. Type: Brazil, Minas Gerais, Barra do Caioa, St.-Hilaire 1442 (lectotype P, isolectotype F (fragment)).

Erect shrubs to 2 m tall; young branches green, increasingly golden brown lanose distally. Leaves of major branches broadly lanceolate to ovate, 20-35 x 5-15 mm, including a 3-7 mm long petiole, mucronate with the apicule less than 1 mm, densely tomentose to lanose on both surfaces; primary vein and sometimes two secondary veins distinct beneath. Racemes robust, lateral and terminal with lateral branches emerging at right angles to main axis; flowering stalks subtended by bracts shorter than 5 mm, separated distally into peduncle and pedicel by a pair of linear, up to 2 mm long bracteoles; sepals pink to deep crimson, turning white with age, lanose-strigose dorsally; glandular petals deep pink to red, petaloid petals red above, green below, 5-10 mm long, variably connate at base for 0.10-0.35 mm; stamens (3-)4, inserted at base of ovary with the petaloid petals, longer pair 5.0-5.4 mm. Fruit body green, 4.5-6.5 mm in diam. excluding spines, densely tomentose, spines red or yellow apically, 2.5-5.0 mm long, 0.25 mm or less in diam. at base with dense barbs along shaft and a very few, long, unicellular trichomes at base.

Distribution: Found in sandy areas and littoral dunes of E Brazil from Amazonas to Minas Gerais at elevations from sea level to 900 m. One specimen has been collected from E Bolivia. It is possible that the species will eventually be found in the flora area. Some specimens from Guyana belonging to *K. ixine* have been mistakenly referred to this species.

Phenology: Flowering occurs sporadically throughout the year across its range.

WOOD AND TIMBER

VOCHYSIACEAE

by

PIERRE DÉTIENNE[5] AND BEN J.H. TER WELLE[6]

WOOD ANATOMY

FAMILY CHARACTERISTICS

Vessels diffuse, solitary and in radial multiples, perforations simple, intervascular pits alternate, vestured. Vessel-ray and vessel-parenchyma pits similar to the intervascular pits except in *Erisma*.
Rays uniseriate, numerous, less numerous in *Qualea* and *Ruizterania*, and 2-6-seriate, weakly heterogeneous.
Parenchyma paratracheal aliform to strongly confluent; apotracheal banded parenchyma in *Erisma*.
Fibres non-septate in *Ruizterania*, *Salvertia* and *Vochysia*, and septate and non-septate in *Erisma* and *Qualea*, with simple to minutely bordered pits.

Special characteristics

Included phloem embedded in parenchyma bands, and in some rays of *Erisma*.

Silica grains are found in the ray cells of some species of *Erisma*, *Qualea*, *Salvertia*, and *Ruizterania*. In some of these species silica grains also occur in the parenchyma. Vitreous/amorphic silica occasionally present in some samples as well.

Traumatic longitudinal intercellular canals common in *Qualea*, *Salvertia*, and *Vochysia*, not observed in *Erisma* and *Ruizterania*.

All the woods have a positive reaction (blue color) on treatment with a solution of Chrome-Azurol-S.

[5] C.I.R.A.D.-Forêt, Maison de la Technologie, B.P. 5035, Montpellier, Cedex 1, France.

[6] Herbarium Division, Department of Plant Ecology and Evolutionary Biology, Heidelberglaan 2, 3584 CS Utrecht, The Netherlands.

KEY TO THE GENERA

1 Parenchyma banded; included phloem strands present · · · · · · · · · · *Erisma*
Parenchyma predominantly paratracheal; included phloem absent · · · · · 2

2 Rays of two types, uniseriates predominant · 3
Rays of two types, multiseriate predominant · 4

3 Rays more than 9 per mm · · · · · · · · · · · · · · · · · *Salvertia convallariodora*
Rays less than 9 per mm · *Vochysia*

4 Rays 4-6-seriate, without silica grains · · · · · · · · · · · · · · · · · *Qualea dinizii*
Rays 2-4-seriate, often with silica grains · 5

5 Chambered parenchyma cells present, with or without rhombic crystals; wood pinkish brown · *Qualea*
Chambered parenchyma cells absent; wood grayish brown · · · *Ruizterania*

ALTERNATIVE KEY

1 Parenchyma banded; included phloem strands present · · · · · · · · · · *Erisma*
Parenchyma predominantly paratracheal; included phloem absent · · · · · 2

2 Chambered parenchyma cells present, with or without rhombic crystals · · · 3
Chambered parenchyma cells absent · 4

3 Uniseriate rays relatively few, multiseriate rays 2-3(4) cells wide, marginal rows of square and upright cells 0-1, number of rays 4-6 per mm; 3-6 vessels per sq. mm · *Qualea*
Uniseriate rays relatively few, multiseriate rays 2-3(-4) cells wide, marginal rows of square and upright cells 0-1, number of rays 3-5 per mm; 8-12 vessels per sq. mm · *Qualea dinizii*
Uniseriate rays relatively numerous, multiseriate rays 3-5 cells wide, marginal rows of square and upright cells 1-5, number of rays 10-12 per mm; 4-5 vessels per sq. mm · · · · · · · · · · · · · · · *Salvertia convallariodora*

4 Number of vessels per sq. mm 4-5; uniseriate rays relatively few, multiseriate rays 2-3 cells wide, marginal rows of square and upright cells 0-1; traumatic canals absent · *Ruizterania*
Number of vessels per sq. mm 1-4; uniseriate rays relatively numerous, multiseriate rays 3-6(8) cells wide, marginal rows of square and upright cells 1-6; traumatic canals present · *Vochysia*

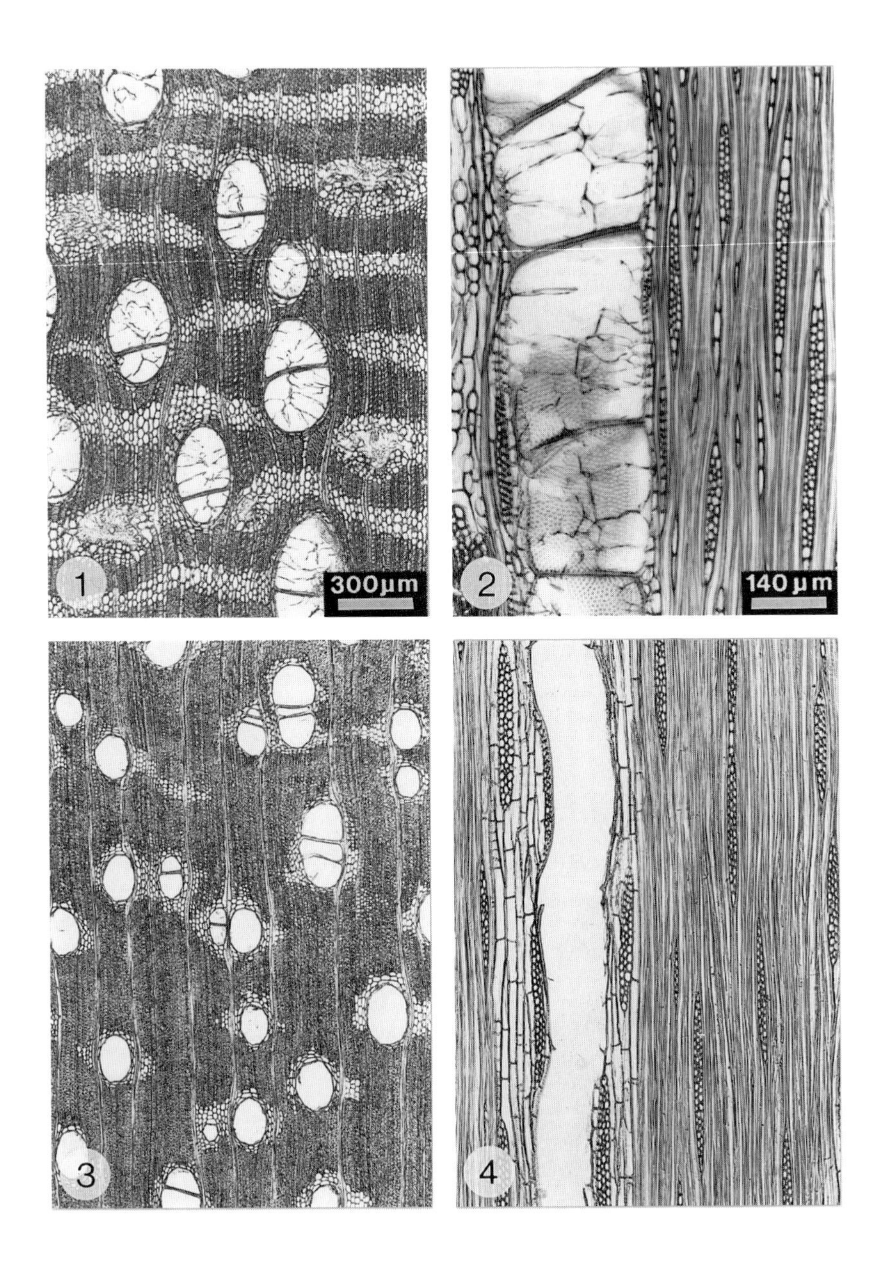

Fig. 1. Transverse section. *Erisma uncinatum* Warm.
Fig. 2. Tangential section. *Erisma uncinatum* Warm.
Fig. 3. Transverse section. *Qualea coerulea* Aubl.
Fig. 4. Tangential section. *Qualea coerulea* Aubl.

GENERIC DESCRIPTIONS

ERISMA Rudge – Figs. 1, 2.

Vessels diffuse, solitary (28-63%) and in radial multiples of 2-4, round to slightly oval, 1-3 per sq. mm, diameter 210-310(140-430) mm. Vessel-member length: 445-560(355-840) µm. Perforations simple. Intervascular pits alternate, round to slightly polygonal, vestured, 9-11 µm, sometimes with coalescent apertures. Vessel-ray and vessel-parenchyma pits simple or with narrow borders, large, round to vertically elongated. Thin-walled tyloses very common.
Rays in two types, 7-8 per mm, uniseriate, up to 145-260 µm (= 2-9 cells) high, composed of square and upright cells, 2-3(-4)-seriate, up to 300-515 µm (= 14-22 cells) high, heterogeneous, composed of procumbent cells and 1-6 marginal rows of square and upright cells. Silica grains sometimes present, rhombic crystals very scarce.
Parenchyma abundant, in concentric bands 3-6(-10) cells wide, 3-4 bands per mm. Strands of 4(-6) cells. Inflated cells rare to common in some samples. Small silica grains sometimes present, rhombic crystals very scarce.
Fibres septate, and a few non-septate, lumen up to 11-20 mm, walls up to 4-8 µm. Pits simple, small, restricted to the radial walls. Length: 1250-1500(920-1825) µm. F/V ratio: 2.3-3.0.
Included phloem diffuse, associated with parenchyma bands, and common, also present in a few rays.

Studied: *E. floribundum*, *E. nitidum*, *E. uncinatum*

QUALEA Aubl. – Figs. 3, 4.

Vessels diffuse, solitary (38-76%) and in radial multiples of 2-3(-5), round to slightly oval, 3-6 per sq. mm (8-12 in *Q. dinizii*), diameter 135-235(95-305) µm. Vessel-member length: 430-725(245-1010) µm. Perforations simple. Intervascular pits alternate, round to slightly polygonal, vestured, 6-7.5 µm, sometimes with coalescent apertures. Vessel-ray and vessel-parenchyma pits similar to the intervascular pits. Tyloses and white deposits sometimes present.
Rays in two types, 4-6(3-8) per mm, few uniseriates, up to 175-285 µm (= 6-14 cells) high, heterogeneous, composed of procumbent and square cells, and 2-3(-4)-seriate, up to 460-805 µm (= 25-51 cells) high, weakly heterogeneous, composed of procumbent cells, and 0-1 marginal row of square cells, sometimes up to 6 marginal rows. Silica grains absent or present, variable in size and shape.

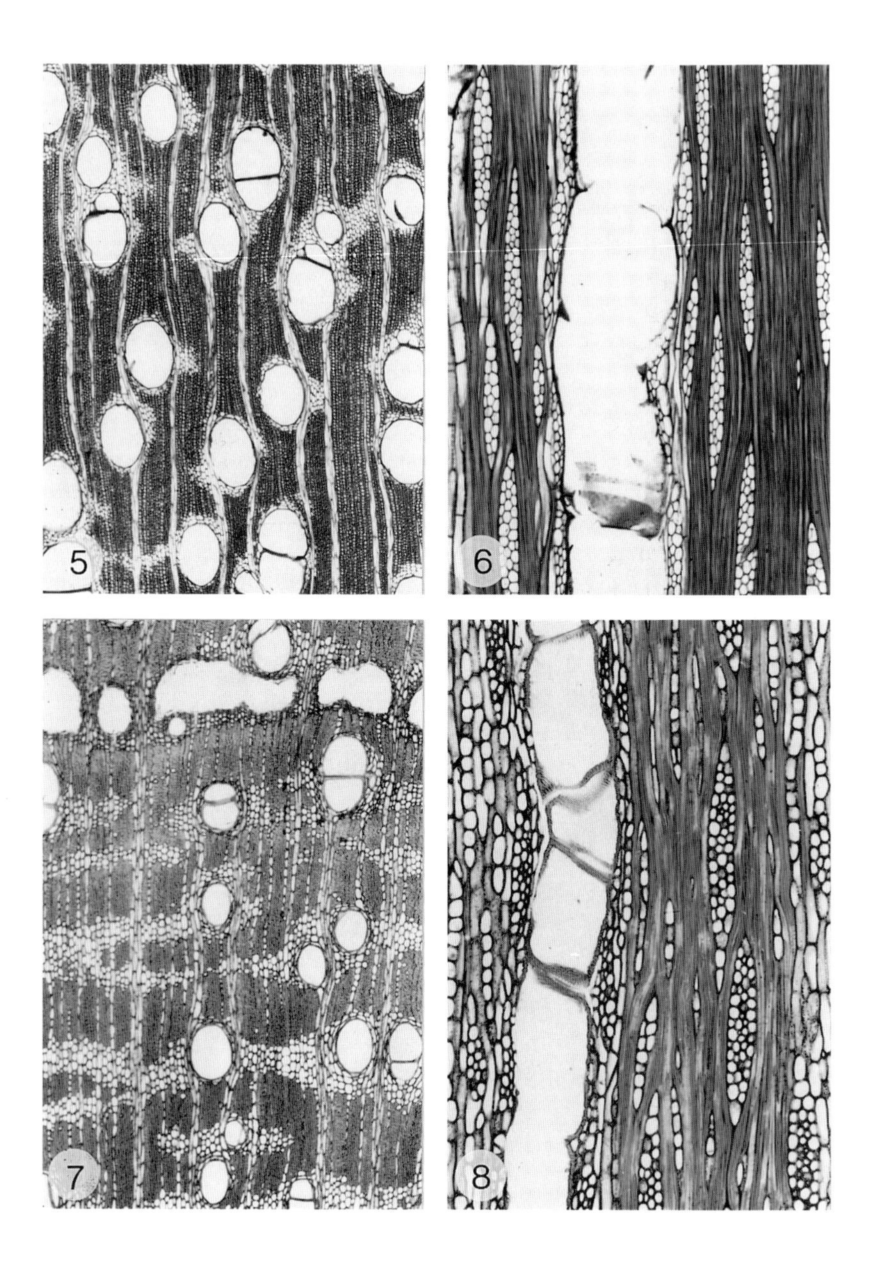

Fig. 5. Transverse section. *Ruizterania albiflora* Marc.-Berti
Fig. 6. Tangential section. *Ruizterania albiflora* Marc.-Berti
Fig. 7. Transverse section. *Salvertia convallariodora* A.St.-Hil.
Band of traumatic canals.
Fig. 8. Tangential section. *Salvertia convallariodora* A.St.-Hil.

Parenchyma paratracheal lozenge- to winged-aliform, rarely to strongly confluent, apotracheal diffuse rare, and occasionally in short bands including traumatic canals. Strands of (2-)4(-8) cells. Rhombic crystals in chambered cells, 2-12(-16) per strand, generally present.
Fibres non-septate to almost all septate, lumen up to 7-17 µm, walls up to 5-7 µm. Pits simple, confined to the radial walls, 3-5 µm. Length: 1260-1635(690-2030) µm. F/V ratio: 1.8-3.1.

Studied: *Q. acuminata, Q. coerulea, Q. dinizii, Q. paraensis, Q. rosea, Q. tricolor*

RUIZTERANIA Marc.-Berti – Figs. 5, 6.

Vessels diffuse, solitary (30-68%) and in radial multiples of 2-3(-5), round to slightly oval, 4-5(3-7) per sq. mm, diameter 160-235(150-290) µm. Vessel-member length: 445-650(270-835) µm. Perforations simple. Intervascular pits alternate, round, vestured, 7-8 µm. Vessel-ray and vessel-parenchyma pits similar to the intervascular pits. White deposits common.
Rays in two types, 4-6(3-8) per mm, few uniseriates, up to 170-345 µm (= 6-11 cells) high, heterogeneous, composed of procumbent and some square and upright cells, and 2-3-seriate, up to 625-725 µm (= 20-39 cells) high, weakly heterogeneous, composed of procumbent cells, and margins of 0-1(2-3) rows of square and/or upright cells. Silica grains round, 5-15 µm, but also other shapes, solid to amorphic, present in the ray cells.
Parenchyma paratracheal lozenge- to winged-aliform, occasionally confluent, apotracheal diffuse rare, and in tangential wavy bands. Strands of (2-)4-6 cells. Silica grains rare, rhombic crystals absent or some present.
Fibres non-septate, lumen up to 11-16 µm, walls up to 5-7 µm. Pits simple, small, confined to the radial walls, 5-7 µm. Length: 1375-1500(840-2115) µm. F/V ratio: 2.3-2.9.

Studied: *R. albiflora, R. ferruginea*

SALVERTIA A.St.-Hil. – Figs. 7, 8.

Vessels diffuse, solitary (25-54%) and in radial multiples of 2-4, and irregular clusters of 3-6, round to slightly oval, 3-4(2-6) per sq. mm, diameter 175-190(115-220) µm. Vessel-member length: 465-515(280-770) µm. Perforations simple. Intervascular pits alternate, round or polygonal, vestured, 6-7 µm. Vessel-ray and vessel-parenchyma pits similar to the intervascular pits or slightly larger. Tyloses present, sometimes with vitreous silica. White deposits sometimes present.

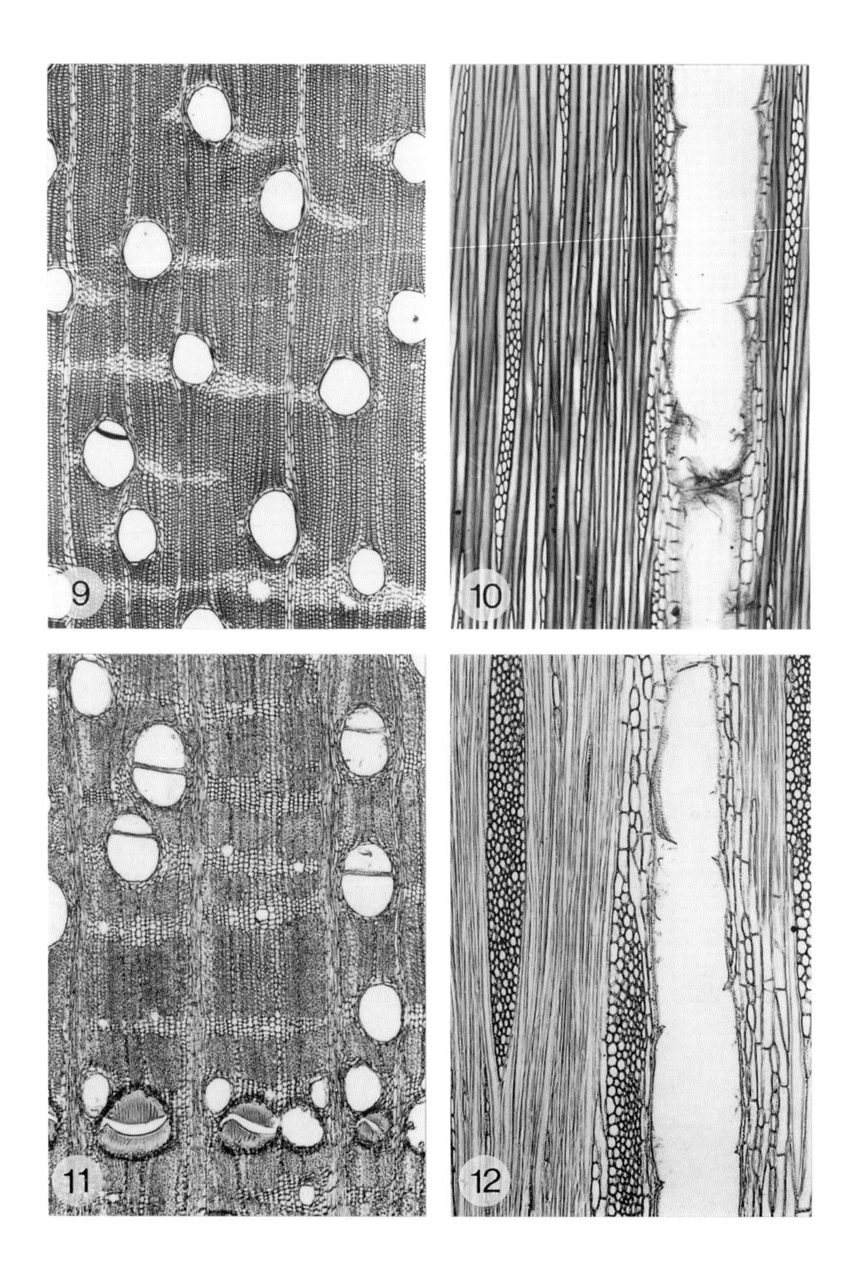

Fig. 9. Transverse section. *Vochysia densiflora* Spruce ex Warm.
Fig. 10. Tangential section. *Vochysia densiflora* Spruce ex Warm.
Fig. 11. Transverse section. *Vochysia guianensis* Aubl.
Band of traumatic canals and inflated parenchyma cells.
Fig. 12. Tangential section. *Vochysia guianensis* Aubl.

Rays in two types, 10-12 per mm, uniseriates, up to 215 μm (= 2-12 cells) high, heterogeneous, composed of square and upright cells, and 3-5-seriate, up to 540-850 μm (= 35 cells) high, heterogeneous, composed of procumbent cells with 1-5 marginal rows of square and upright cells. Rather amorphic silica grains occasionally present, small.
Parenchyma paratracheal winged-aliform, strongly confluent or in irregular concentric bands, 2-8 cells wide. Strands of 3-5 cells. Some rhombic crystals in chambered cells.
Fibres non-septate, lumen up to 6-12 μm, walls up to 5-8 μm. Pits minutely bordered, more frequent on the radial than on the tangential walls, 5-7 μm. Length: 1170-1445(720-1570) μm. F/V ratio: 2.5-2.8.
Traumatic longitudinal canals, large and small in diameter, in short tangential bands and embedded in parenchyma, present.

Studied: *S. convallariodora*

VOCHYSIA Aubl. – Figs. 9, 10, 11, 12.

Vessels diffuse, solitary (58-89%) and in radial multiples of 2-3, and occasionaly in clusters of 3-6, round to slightly oval, 1-4(0-7) per sq. mm, diameter 180-300(115-330) μm. Vessel-member length: 545-700(330-975) μm. Perforations simple. Intervascular pits alternate, round, vestured, 7-9 μm. Vessel-ray and vessel-parenchyma pits similar to the intervascular pits. Tyloses, if present, thin-walled. White deposits sometimes present.
Rays in two types, 5-8 per mm, uniseriate, up to 195-335 μm (= 2-12 cells) high, heterogeneous, composed of square and upright cells, and (2-)3-6(-8)-seriate, 575-2000(2800) μm (= 34-100 cells) high, heterogeneous, composed of procumbent cells with 1-6 marginal rows of square and upright cells. Sheath cells scarce.
Parenchyma paratracheal winged-aliform, occasionally to strongly confluent in wide tangential bands, apotracheal parenchyma sometimes present in very short bands or in long bands including traumatic cells. Strands of (2-)4-6 cells. Inflated cells, vertically in rows of 2 or 4, observed in *V. guianensis* and *V. surinamensis*.
Fibres non-septate, lumen up to 12-32 μm, walls up to 4-8 μm. Pits simple to minutely bordered, frequent on the radial walls, rare on the tangential walls, 4-5 μm. Length: 1245-1700(785-3240) μm. F/V ratio: 1.9-3.2.

Studied: *V. cayennensis*, *V. densiflora*, *V. ferruginea*, *V. glaberrima*, *V. guianensis*, *V. neyratii*, *V. schomburgkii*, *V. speciosa*, *V. surinamensis*, *V. tetraphylla*, *V. tomentosa*

Erisma

Botanical	*E. uncinatum* is the only timber producing species.
Tree	Dominant tree, up to 35 m high, with thick buttresses only well developed (2-3 high) in biggest trees. Diameter 50-100(-150) cm. Bole cylindrical, 20-25 m long.
Description of the wood	Sapwood grayish white, 4-12 cm wide, well demarcated from the pinkish to purplish brown heartwood. Texture coarse. Grain straight. Lustre medium.
Weight	Specific gravity 450-650 kg per cubic metre (12%).
Shrinkage	From green to ovendry: radial 4.7% (3-7); tangential 10.5% (8-14); volumetric 16% (12-21).
Seasoning properties	Easy and quick to dry.
Mechanical properties	Crushing strength: 500(420-620) kg/sq. cm. Static bending: 1310(1080-1480) kg/sq. cm. Modulus of elasticity: 135(105-175) kg/sq. cm x 1000.
Working properties	Works easily.
Durability	Resistant to decay but susceptible to termite attack.
Preservation	Resistant to impregnation.
Uses	Interior and exterior construction, joinery, furniture.
Supply	Relatively abundant in FG, less in SU, rare in GU.
Trade names	Singri kwari (SU). Felli kouali, Jaboty, Manonti kouali (FG).

Qualea

Botanical	*Q. coerulea* and *Q. rosea* are the two main timber producing species.
Tree	Dominant tree, up to 30 m high, basally swollen. Diameter 50-80(-100) cm. Bole cylindrical, 15-20 m long.
Description of the wood	Sapwood grayish white, 3-5 cm wide, demarcated from the reddish brown heartwood. (Heartwood gray-brown in *Q. dinizii*). Texture medium. Grain slightly interlocked. Lustre high.
Weight	Specific gravity 580-800 kg per cubic metre (12%).
Shrinkage	From green to ovendry: radial 6% (5-7.5); tangential 10.5% (9.5-12.5); volumetric 18% (15-21).
Seasoning properties	Air seasons easily to moderately difficult.
Mechanical properties	Crushing strength: 680(600-830) kg/sq. cm. Static bending: 1630(1200-2100) kg/sq. cm. Modulus of elasticity: 155(115-200) kg/sq. cm x 1000.
Working properties	Moderately difficult to work, but without problem with specially hardened cutters.
Durability	Moderately resistant to decay and to termite attack.
Preservation	Rather good with pressure-vacuum treatments.
Uses	Interior joinery, carpentry, flooring. Possible in exterior work, but without contact with the soil.
Supply	Generally *Qualea* and *Ruizterania* are logged together. Average volume of trees over 40 cm diameter: 2-3.2 cubic metre per hectare (FG).

Trade names	Manaw, Yakopi (GU). Berggronfoeloe, Gronfoeloe (SU). Gonfolo, Gonfolo kouali, Gonfolo rose (FG).

Ruizterania

Botanical	*R. albiflora* is recorded as timber producing species.
Tree	Dominant tree, up to 30 m high, with thick and short (1-2 m high) buttresses.
Description of the wood	Sapwood cream-coloured, more or less demarcated from the pale grayish (sometimes with a pinkish shade) brown heartwood. Texture medium. Interlocked grain frequent, but rarely well developed. Lustre medium.
Weight	Specific gravity 650-750(-850) kg per cubic metre (12%).
Shrinkage	From green to ovendry: radial 6% (5-7); tangential 10.5% (9-12); volumetric 16.5% (13-19).
Seasoning properties	Air seasons easily to moderately difficult.
Mechanical properties	Crushing strength: 670(610-850) kg/sq. cm. Static bending: 1550(1430-1750) kg/sq. cm. Modulus of elasticity: 135(120-170) kg/sq. cm x 1000.
Working properties	Moderately difficult to work, but without problem with specially hardened cutters.
Durability	Moderately resistant to decay and to termite attack.
Preservation	Rather good with pressure-vacuum treatments.
Uses	Interior joinery, carpentry, flooring. Possible in exterior work but without contact with the soil.

Supply	Generally *Ruizterania* and *Qualea* are logged together. Average volume of trees over 40 cm diameter: 2-3.2 cubic metre per hectare (FG).
Trade names	Manau, Yakopi (GU). Hoogland gronfoeloe (SU). Gonfolo, Gonfolo gris (FG).

Salvertia

Description of the wood	Sapwood yellowish, demarcated from the pinkish brown heartwood. Texture coarse. Grain straight. Lustre medium.
Weight	Specific gravity 590-700 kg per cubic metre (12%).

Vochysia

Botanical	The following species are recorded as timber producing species: *V. cayennensis*, *V. densiflora*, *V. guianensis*, *V. neyratii*, *V. speciosa*, *V. surinamensis*, *V. tetraphylla* and *V. tomentosa*.
Tree	Dominant tree, up to 35 m high, basally swollen? never up to 1 m high. Diameter 50-100 cm. Bole straight cylindrical, 10-25 m long.
Description of the wood	Sapwood 2-5 cm wide, cream-coloured to pale beige, poorly demarcated from the pinkish pale brown heartwood. Texture medium to coarse. Grain straight or somewhat interlocked. Lustre high.
Weight	Specific gravity generally 450-650 kg per cubic metre (12%), sometimes less 350-500 in *V. cayennensis*, *V. tomentosa* or more 550-700 in *V. guianensis*, *V. surinamensis*.
Shrinkage	From green to ovendry: radial 3-5%; tangential 9-14%; volumetric 13-20%.

Seasoning properties	Air seasons rapidly to slowly, prone to warp with some checking. Collaps possible in thick boards.
Mechanical properties	Crushing strength: 350-530 kg/sq. cm. Static bending variable according to species and density: 700-1400 kg/sq. cm. Modulus of elasticity: 75-140 kg/sq. cm x 1000.
Working properties	Works well to excellent except when tension wood is abundant (rough surface).
Durability	Poorly resistant to decay and susceptible to termite attack.
Preservation	Can be treated with a medium penetration and absorbtion of preservatives.
Uses	Interior joinery and furniture. Possible for interior or plywood.
Supply	Average volume of trees over 40 cm in diameter: 0.5-2 cubic metre per hectare in FG.
Trade names	Deokunud, Hill iteballi, Iteballi, Kwaru (GU). Appel kwarie, Wana kwarie, Watra kwarie, Wiswis kwarie (SU). Achiwa, Grignon fou, Grignon Sainte Marie, Kouali, Koupi kouali, Moutende kouali, Papakai kouali, Wachi-wachi-kouali (FG).

LITERATURE ON WOOD AND TIMBER

Anonymus. 1988. Madeiras de Amazônia. Caracteristicas e utilizacao. I.B.D.F.F., Brasilia.

Anonymus. 1989. Bois des DOM-TOM. Tome I. Guyane. CIRAD-Forêt, Nogent-sur-Marne.

Anonymus. 1990. Atlas des bois tropicaux d'Amérique latine. A.T.I.B.T., Nogent-sur-Marne.

Bena, P. 1960. Essences forestières de Guyane. Bureau Agricole et Forestier Guyanais. Imprimerie Nationale, Paris.

Berni, C.A., E. Bolza & F.J. Christensen. 1979. South American Timbers. The characteristics, properties and uses of 190 species. C.S.I.R.O., Melbourne.

Chudnoff, M. 1984. Tropical Timbers of the World. Agriculture Handbook No. 607. US Dept. of Agriculture, Forest Service, Madison.

Détienne, P., P. Jacquet & A. Mariaux. 1982. Manuel d'identification des bois tropicaux. Tome 3. Guyane française. C.T.F.T., Nogent-sur-Marne.

Détienne, P. & P. Jacquet. 1983. Atlas d'identification des bois de l'Amazonie et des régions voisines. C.T.F.T., Nogent-sur-Marne.

Fanshawe, D.B. 1948. Forest Products of British Guiana. Part 1. Principal Timbers. Forestry Bulletin 1 (New Series).

Japing, C.H. & H.W. Japing. 1960. Houthandboek Surinaamse houtsoorten. Dienst 's Lands Bosbeheer, Paramaribo.

Mennega, A.M.W. 1948. Surinam Timbers I. General introduction, Guttiferae, Vochysiaceae, Anacardiaceae, Icacinaceae. Nat. Wet. Studiekring Suriname en Curaçao. No. 3. Marinus Nijhoff, Den Haag.

Lindeman, J.C. & A.M.W. Mennega. 1963. Bomenboek voor Suriname. Dienst 's Lands Bosbeheer, Paramaribo.

Mennega, E.A., W.C.M. Tammens-de Rooij & M.J. Jansen-Jacobs (Eds.). 1988. Checklist of woody plants of Guyana. Tropenbos Technical Series 2. Tropenbos Foundation, Ede.

Metcalfe, C.R. & L. Chalk. 1950. Anatomy of the Dicotyledons. Volume 1. Clarendon Press, Oxford.

Normand, D. 1966. Les Kouali, Vochysiacées de Guyane, et leurs bois (1). Bois et Forêts des Tropiques 110: 3-11.

Normand, D. 1967. Les Kouali, Vochysiacées de Guyane, et leurs bois (2). Bois et Forêts des Tropiques 111: 5-17.

Pfeiffer, J.Ph. 1926. De houtsoorten van Suriname. Deel I. De Bussy, Amsterdam.

Polak, A.M. 1992. Major Timber Trees of Guyana. A field Guide. Tropenbos Series 2. The Tropenbos Foundation, Wageningen.

Quirk, J.T. 1980. Wood anatomy of the Vochysiaceae. I.A.W.A. Bull. n.s. 1(4): 172-179.
Record, S.J. & R.W. Hess. 1943. Timbers of the New World. Yale University Press, New Haven, CT.
Vink, A.T. 1965. Surinam timbers. Surinam Forest Service, Paramaribo.

EUPHRONIACEAE

by

NARCISANA ESPINOZA DE PERNÍA[7] AND BEN J.H. TER WELLE[8]

WOOD ANATOMY

GENERIC DESCRIPTION

EUPHRONIA Mart. – Figs. 13, 14, 15, 16.

Vessels diffuse, solitary (up to 100%) and only very occasionally an irregular cluster of 2, round, sometimes oval, occasionally angular, 8-13(3-15) per sq. mm, diameter 90-120(58-150) µm. Vessel-member length: 530-732(323-1029) µm. Perforations simple. Intervascular pits very scarce, alternate, round, 1-3 µm. Vessel-ray and vessel-parenchyma pits similar to the intervascular pits.
Rays uniseriate, and mostly biseriate (at least in part), 12(10-14) per mm, up to 680, sometimes up to 800 µm (=15-24 cells) high, heterogeneous, composed of square, upright and slightly procumbent cells, biseriate rays often with uniseriate margins of square and upright cells.
Parenchyma conspicuous, paratracheal winged- to lozenge-aliform, and mostly aliform-confluent, sometimes in more or less continuous wavy bands. Strands of 2-8 cells. Few rhombic crystals present, sometimes crystals absent.
Fibres non-septate, thick-walled, lumen up to 6-14 µm, walls up to 8-10 µm. Pits bordered, abundant on radial and tangential walls, 5 µm. Length: 961-1226(764-1382) µm. F/V ratio: 1.67-1.87.

Studied: *E. guianensis, E. acuminatissima* (both from Venezuela)

[7] Universidad de Los Andes, Facultad de Cienciás Forestales, Laboratorio de Anatomía de Maderas, Avenida Los Chorros de Milla, Mérida, Venezuela.

[8] Herbarium Division, Department of Plant Ecology and Evolutionary Biology, Heidelberglaan 2, 3584 CS Utrecht, The Netherlands.

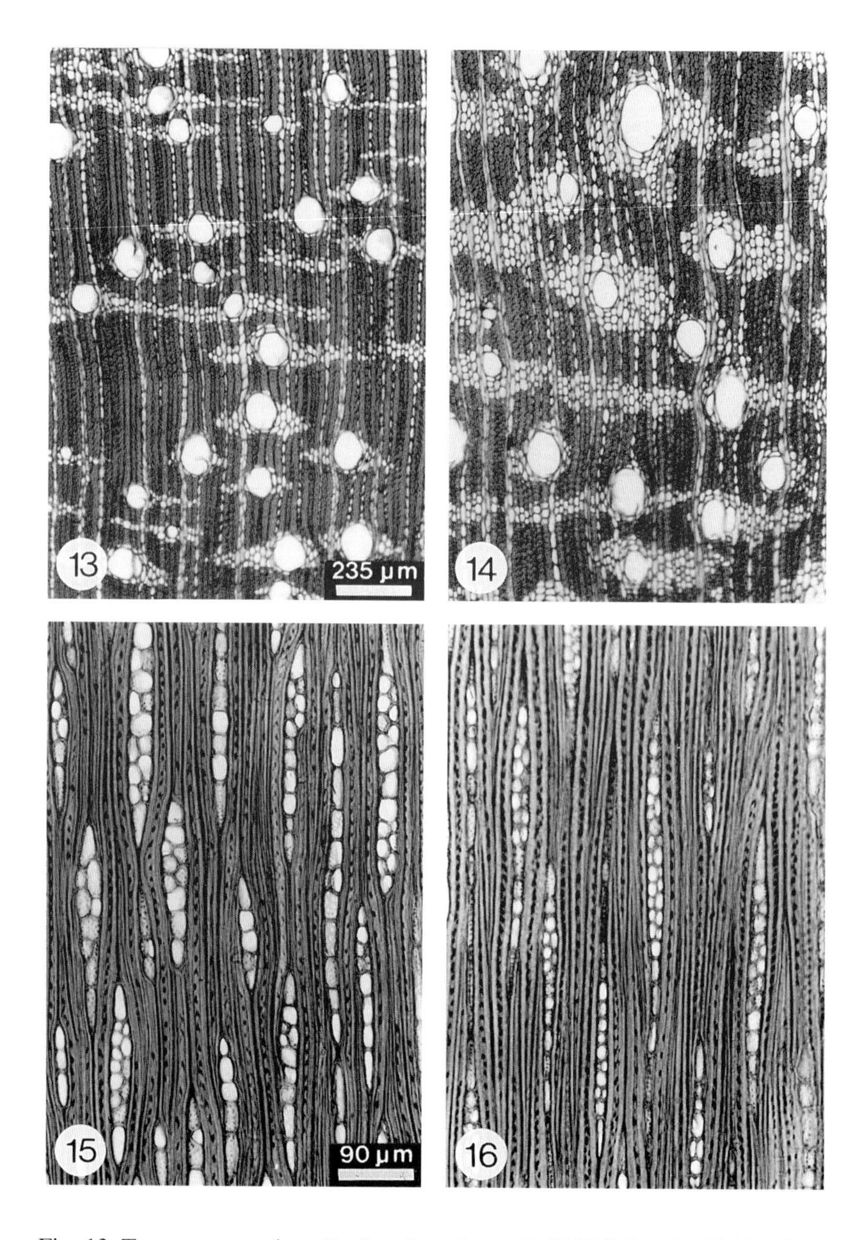

Fig. 13. Transverse section. *Euphronia guianensis* (R.H.Schomb.) Hallier f.
Fig. 14. Transverse section. *Euphronia guianensis* (R.H.Schomb.) Hallier f.
Fig. 15. Tangential section. *Euphronia guianensis* (R.H.Schomb.) Hallier f.
Fig. 16. Tangential section. *Euphronia acuminatissima* Steyerm.

TIMBERS AND THEIR PROPERTIES

Euphronia

Description of the wood	No difference between sapwood and heartwood, brownish. Texture medium. Grain straight. Lustre medium.
Weight	Specific gravity ca. 500-700 kg per cubic metre (12%).

LITERATURE ON WOOD AND TIMBER

Espinoza de Pernía, N. 1989. Estudio xilologico del genero Euphronia. Pittieria 18: 57-58.

TRIGONIACEAE

by

BEN J.H. TER WELLE[9] AND PIERRE DÉTIENNE[10]

WOOD ANATOMY

GENERIC DESCRIPTION

TRIGONIA Aubl. – Figs. 17, 18, 19, 20, 21.

Growth rings absent, or faint and scarce. When present due to a slight difference in vessel diameter and occasionally tangentially flattened fibres as observed in the transverse sections.
Vessels diffuse, solitary (90-95%) and some multiples of 2(-3), oval and sometimes round, 6-45(4-51) per sq. mm, diameter variable per sample, in two size groups, 115-230(80-280) µm and 45-80(25-115) µm respectively. Vessel-member length: 403-584(253-725) µm. Perforations simple. Intervascular pits very scarce, alternate, round, 1-3 µm. Vessel-ray and vessel-parenchyma pits alternate, round, 2-3 µm.
Rays uniseriate and 2-3(-4)-seriate, and 4-8-seriate in *T. laevis* var. *microcarpa*, 6-10(5-11) per mm, uniseriates up to 770-1300 µm (= 15-31 cells) high, heterogeneous, composed of square and some upright cells, multiseriates up to 1200-2200(-3500) µm (= 42-160 cells) high, heterogeneous, composed of procumbent cells, and up to 2-6(-9) uniseriate marginal rows of square cells.
Parenchyma scanty paratracheal to vasicentric, in irregular apotracheal tangential wavy and short bands of 2-4 cells wide, and diffuse. Strands of 3-8 cells. A few rhombic crystals present in *T. laevis* var. *microcarpa*.
Fibres non-septate, lumen up to 13-18 µm, walls up to 2-7 µm. Pits bordered, abundant on radial and tangential walls, 2-4 mm in diameter. Length: 708-1117(587-1277) µm. F/V ratio: 1.76-2.53.

Note: Spiral thickenings observed in some ray and parenchyma cells of *T. nivea* var. *nivea*.

Studied: *T. candelabra, T. coppenamensis, T. hypoleuca, T. laevis* var. *microcarpa, T. nivea* var. *nivea*

[9] Herbarium Division, Department of Plant Ecology and Evolutionary Biology, Heidelberglaan 2, 3584 CS Utrecht, The Netherlands

[10] C.I.R.A.D.-Forêt, Maison de la Technologie, B.P. 5035, Montpellier, Cedex 1, France

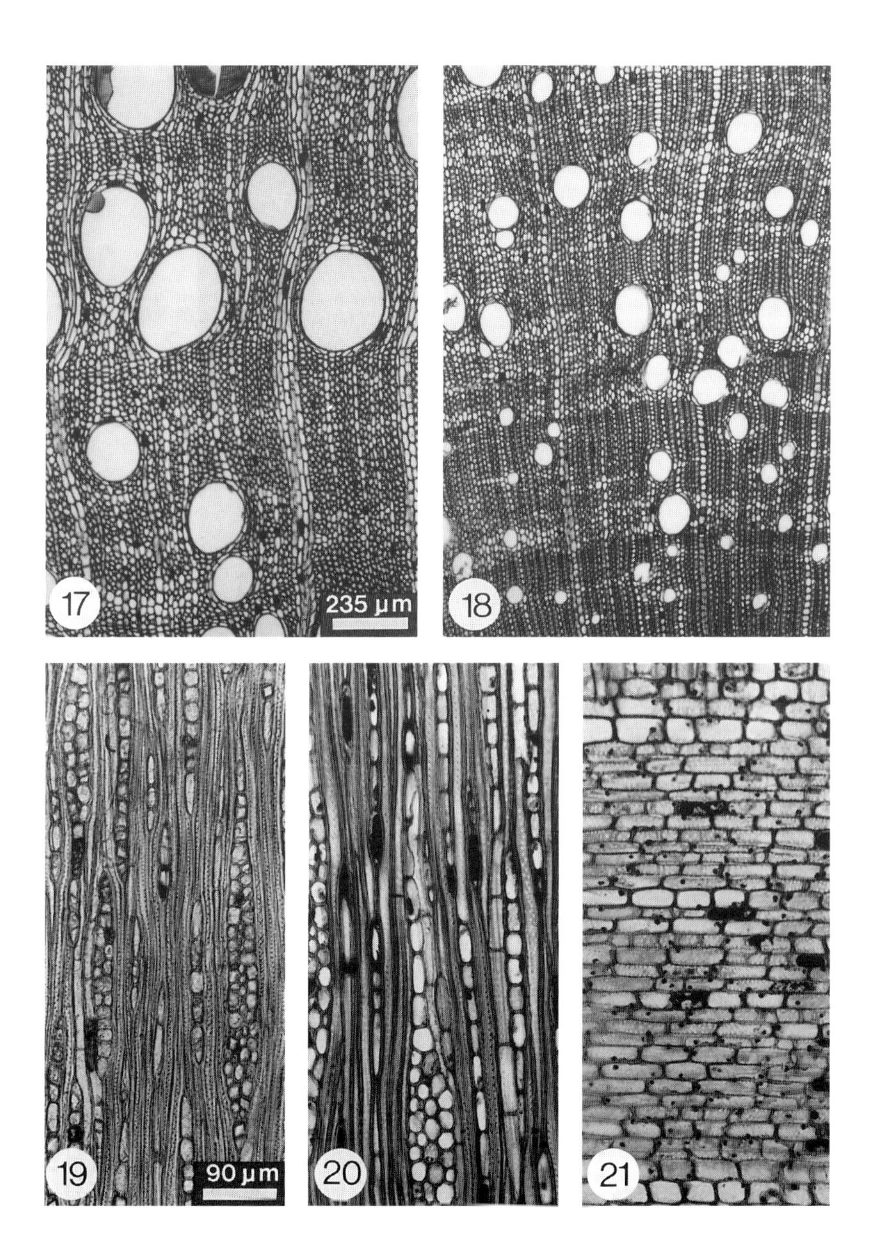

Fig. 17. Transverse section. *Trigonia coppenamensis* Stafleu
Fig. 18. Transverse section. *Trigonia candelabra* Lleras
Fig. 19. Tangential section. *Trigonia laevis* Aubl. var. *microcarpa* Sagot
Fig. 20. Tangential section. *Trigonia coppenamensis* Stafleu
Fig. 21. Radial section. *Trigonia coppenamensis* Stafleu

TIMBERS AND THEIR PROPERTIES

As far as known the wood is of no commercial value.

LITERATURE ON WOOD AND TIMBER

Heimsch, C. 1942. Comparative anatomy of the secondary xylem in the "Gruinales" and "Terebinthales" of Wettstein with reference to taxonomic grouping. Lilloa 8: 83-198.

Metcalfe, C.R. & L. Chalk. 1950. Anatomy of the Dicotyledons. Volume 1. Clarendon Press, Oxford.

NUMERICAL LIST OF ACCEPTED TAXA

Vochysiaceae

1. Erisma Rudge
   1-1. E. floribundum Rudge var. floribundum
   1-2. E. nitidum DC.
   1-3. E. uncinatum Warm.

2. Qualea Aubl.
   2-1. Q. acuminata Spruce ex Warm.
   2-2. Q. coerulea Aubl.
   2-3. Q. dinizii Ducke
   2-4. Q. grandiflora Mart.
   2-5. Q. mori-boomii Marc.-Berti
   2-6. Q. paraensis Ducke
   2-7. Q. polychroma Stafleu
   2-8. Q. psidiifolia Spruce ex Warm.
   2-9. Q. rosea Aubl.
   2-10. Q. schomburgkiana Warm.
   2-11. Q. tricolor Benoist

3. Ruizterania Marc.-Berti
   3-1. R. albiflora (Warm.) Marc.-Berti
   3-2. R. ferruginea (Steyerm.) Marc.-Berti
   3-3. R. rigida (Stafleu) Marc.-Berti

4. Salvertia A.St.-Hil.
   4-1. S. convallariodora A.St.-Hil.

5. Vochysia Aubl.
   5-1. V. cayennensis Warm.
   5-2. V. costata Warm.
   5-3. V. crassifolia Warm.
   5-4. V. densiflora Spruce ex Warm.
   5-5. V. ferruginea Mart.
   5-6. V. glaberrima Warm.
   5-7. V. guianensis Aubl.
   5-8. V. maguirei Marc.-Berti
   5-9. V. neyratii Normand
   5-10. V. sabatieri Marc.-Berti
   5-11. V. schomburgkii Warm.
   5-12. V. speciosa Warm.
   5-13. V. surinamensis Stafleu var. surinamensis

5-14. V. tetraphylla (G.Mey.) DC.
5-15. V. tomentosa (G.Mey.) DC.
5-16. V. tucanorum Mart. var. tucanorum

Euphroniaceae

1. Euphronia Mart.
   1-1. E. guianensis (R.H.Schomb.) Hallier f.

Trigoniaceae

1. Trigonia Aubl.
   1-1. T. candelabra Lleras
   1-2. T. coppenamensis Stafleu
   1-3. T. hypoleuca Griseb.
   1-4. T. laevis Aubl.
   1-4a. var. laevis
   1-4b. var. microcarpa Sagot
   1-5. T. nivea Cambess. var. nivea
   1-6. T. reticulata Lleras
   1-7. T. subcymosa Benth.
   1-8. T. villosa Aubl.
   1-8a. var. macrocarpa (Benth.) Lleras
   1-8b. var. villosa

Krameriaceae

1. Krameria Loefl.
   1-1. K. ixine Loefl.
   1-2. K. spartioides Klotzsch ex O. Berg
   1-3. K. tomentosa A.St.-Hil.

COLLECTIONS STUDIED
(Numbers in italics represent types)

Vochysiaceae

GUYANA

Abraham, A.A., 347 (1-2)
Boom, B., D. Gopaul & P. Taylor, 8421 (2-)
CLAL, 00215 (3-3)
Cook, C.D.K., 244 (5-14)
Cruz, J.S. de la, 2824 (5-14)
Ek, R.C., 600 (3-)
FD (Forest Dept. British Guiana), 106 (5-4); 155 (5-11); 832 (2-6); 2246 (5-3); 2613 (5-11); *2832* (2-7); 3050 (5-13); 3089, 3176 (5-4); 3260 (5-11); 3743 (1-2); 5218 (5-13); 5801, 5929 (1-3); 5934 (5-13); 6771 (2-3); 7643 (5-5)
Hancock, J., s.n. (5-15)
Hoffman, B., 1857 (2-10)
Görts-van Rijn, A.R.A. et al., 310 (5-3); 483 (5-14)
Jansen-Jacobs, M.J. et al., 680, 1353 (5-14); 1623 (5-6); 1700 (3-2); 2326 5-7; 2426 (5-5): 3142, 3759 (2-3)
Jenman, G.S., 453, 970 (5-14); 4276 (5-11); 7439 (1-2); 7802 (5-11)
Maguire, B., 24841 (3-1); *32645* (5-8); 34251 (3-2)
Myers, J., 5662 (2-6); 5701 (5-5)
Pinkus, A., 240 (3-2)
Pipoly, J.J. et al., 7963, 10405 (2-10); 10578 (5-8); 11707, 11713 (5-14)
Schomburgk, Rich., 511 (5-14); 868 (1-1); *974* (5-2); *983* (2-10); 1537 (3-3)
Schomburgk, Rob., 19 (3-3); 584 (2-10); *585* (5-3); *642* (5-6); *902* = Rich. *1360* (5-11); 1047 (2-10)
Smith, A.C., 2258, 2425 (5-6); 2511 (5-5); 2587 (5-14); 2709 (3-1); 2807 (5-5); 3250 (2-3); 3270 (5-14)
Tate, G., 206 (2-10)
Thurn, E.F. im, 79 (2-10)
Tillett, S.S., C.L. Tillett & Boyan, 45030 (3-2); 45222, 45523 (2-10)
Tillett, S.S. & C.L. Tillett, 45727 (3-3); 45836 (2-10); 45837 (3-3)
Wilson-Browne, G., 387 (1-3)

SURINAME

BBS (Boschbeheer Suriname), 11 (2-2); 48 (3-1); 53 (2-3); 82 (5-13); 405, 600, 1011, 1015, 1016 (2-2); 1062 (3-1)
Boon, H., 1051, 1084, 1091 (5-14)
Burger, D., 16 (5-15)
BW (Boschwezen), 360 (5-7); 390 (5-14); 392 (5-7); 437, 448 (5-15); 458 (5-7); 738 (2-3); 1081 (5-4); 1133 (5-7); 1236 (5-15); 1293 (5-4); 1346 (2-2); 1361 (5-7); 1372 (2-2); 1382 (5-7); 1442 (5-4); 1501 (2-3); 1551 (5-4); 1694 (2-2); 1763 (5-2); 1967 (5-7); 2087, 2092 (5-13); 2354 (5-15); 2394, 2814 (2-3); 3260 (5-13); 3342, 3370, 3380 (1-3); 3542

(5-14); 3786 (2-2); 3811 (2-3); 4017 (2-2); 4031 (5-14); 4048 (5-7); 4093 (5-4); 4095, 4109 (5-15); 4173 (2-3); 4217, 4257 (5-15); 4260 (2-9); 4268 (5-2); 4287 (2-2); 4320 (2-3); 4751 (1-3); 4850 (5-15); 4914 (2-3); 5026 (5-14); 5434 (1-3); 5457 (5-14); 5458 (5-7); 5560 (5-14); 5564 (1-3); 5770 (5-7); 6029, 6039 (5-15); 6052 (5-4); 6057 (5-7); 6058, 6061, 6278 (5-15); 6336 (2-9); 6338 (2-2); 6405 (5-15); 6510, *6915* (5-13)

Donselaar, J. van, 1524 (2-2)

Focke, H., 106 (5-14)

Hekking, W.H.A., 1056 (5-14)

Heyde, N.M. & J.C. Lindeman, 177 (5-15)

Irwin, H. et al., 55974 (2-3)

Kappler, A., 1293 (2-2); 1628 (5-14); 2037 (3-1)

Kegel, *684* (5-14)

Kuyper, J., 112 (5-14)

Lanjouw, J., 1210 (5-14)

Lanjouw, J. & J.C. Lindeman, 281 (5-14); 392, 395 (2-9); 539, 2854 (5-4); 2920 (2-9)

LBB (Landsbosbeheer), 10821 (5-6); 11180 (3-1); 11182 (2-9); 11197 (2-3); 12578 (2-2); 14760 (2-9); 16264 (2-3)

Lindeman, J.C., 4879 (2-2); 5074 (5-15); 5208 (5-7)

Lindeman, J.C., A.L. Stoffers et al., 396 (5-14)

Maguire, B., 24917 (5-14)

Maguire, B. et al. 53922 (2-3)

Oldenburger, F.H.F. et al., 229 (4-1); 1158 (2-4)

Procter, J., 4789 (2-2)

Pulle, A., 140 (5-14)

Rombouts, H., 275 (4-1)

Schulz, J.P., 7430 (3-1); 7433 (2-9)

Splitgerber, F., 136 (5-14)

Stahel, G. (Woodherbarium), 46 (1-3); 47 (5-7); 59 (2-3); 59a (2-3); 87 (5-15); 111 (5-4); 374 (5-14)

Versteeg, G., 220, 821 (5-14)

Went, F.C., 535 (5-14)

Wullschlaegel, H., 446 (5-16); 743 (5-14)

FRENCH GUIANA

Aublet, J.B.C.F. d', *s.n.* (2-2); *s.n.* (2-9); *s.n.* (5-7)

BAFOG, 15 M (5-13); 75 M (3-1); 91 M (2-9); 95 M (5-7); 97 M (5-15); 98 M (2-2); 120 M (2-9); 123 M (2-2); 135 M (1-3); 203 M, 273 M (5-14); 309 M (5-13); 312 M (5-14); 314 M (1-3); 315 M (5-14); 316 M (2-2); 317 M (2-9); 318 M (1-3); 320 M (5-15); 322 M (5-7); 333 M (5-4); 1222 (2-2); 1229, 1257 (5-9); 4236, 5112 (5-14); 6088 (2-2); 6143 (2-3); 7054 (2-9); 7056, 7066 (5-7); 7102 (5-15); 7105, 7115 (5-7); 7116 (2-9); 7117, 7118 (5-15); 7119, 7120 (2-9); 7165 (5-7); 7166 (5-15); 7179, 7180 (2-9); 7190 (5-4); 7196, 7197, 7198, 7201 (2-9); 7203, 7205 (2-2); 7246 (5-7); 7286 (5-15); 7376 (2-9); 7377, 7398 (5-7); 7399 (2-9); 7400 (5-15); 7401 (5-15); 7402 (2-9); 7414, 7441 (5-15); 7442 (2-9); 7443 (5-7); 7453 (2-9); 7483 (5-15); 7505 (5-4); 7511 (5-14); 7546 (3-1); 7566 (2-2); 7572 (3-1); 7606

EXTRA GUIANAN

Huber, O., MG *538* (5-7), MG *1844* (2-1) – BRA
Krukoff, B.A., 1401 (1-1) – BRA
Martius, C.F.P. von, *s.n.* (2-4); *s.n.* (5-5); *1179* (5-16) – BRA
Poeppig, E.F., *2633* (1-3) – BRA
Pohl, J.E., *s.n.* (4-1); *s.n.* (5-16) – BRA
Spruce, R., *2612* (2-1); *2627* (5-4) – BRA; *3059* – VEN
Steyermark, J.A., *60914, 90467, 94203* (3-2) – VEN

Euphroniaceae

GUYANA

Schomburgk, Rob. I. Add. Ser. 2, *114* (1-1)

EXTRA GUIANAN

Bernardi, 2679 (VEN; MER, G)
Guevara, J.R., 413 (MER;)
Huber, O., 11289 (U,VEN); 11296 (U, VEN)
Marcano-Berti, L. & Jaime A. Bautista B., 2488 (MER)
Marcano-Berti, L., L. A. Pinto & Ismael Peña S., 125-981, 126-981, 130-981, 138-981 (MER)
Steyermark, J.A. et al., 104162 (VEN)

Trigoniaceae

GUYANA

Fanshawe, D.B., see Forest Dept. British Guiana - F numbers
FD (Forest Dept. British Guiana), 2470 (1-4b); 3036 = F300 (1-3); 4513 = F1777; 6303 = F2973 (1-4b), 7992 (1-6)
Gleason, H.A. et al., 516, 549, 576, 877 (1,3)
Irwin, H.S., 679 (1-5)
Jansen-Jacobs, M.J. et al., 1124 (1-8b)
Jenman, G.S., 1083 (1-8b); 1155, 1296 (1-3); 6736 (1-8b)
Maas, P.J.M. et al., 7124 (1-4a); 7392 (1-5)
Maguire, B. et al., *23192* (1-4a)
Persaud, A.C., 179 (1-4b)
F.V. McConnell & Quelch, J.J., 222 (1-5)
Sandwith, N.Y., 563, 1592 (1-4b)
Schomburgk, Rich., 343 (1-8a); 953 (1-4b)
Schomburgk, Rob., ser. I, *54* earlier sets (1-8a); 54 later sets (1-8b); *56* earlier sets (1-7); 56 later sets (1-3); 63 earlier sets (1-8b); 63 later sets (1-7); ser II 224 = Rich. 315 (1-3) 249 = Rich. 373 (1-7)
Smith, A.C., 2752 (1-4b)
Tillett, S.S. & C.L. Tillett, *45524* (1-6)
Tutin, T.G., 172 (1-4a)

SURINAME

Boon, H., 1069, 1105 (1-3)
BW, 631 (1-4b); 2163 (1-3); 2864 (1-8b); 3289, 6485 = Stahel Wilhelmina Exp., 58 (1-4b)
Donselaar, J. van, 1335 (1-3); *2812*, 2891 (1-1); 2894 (1-4b)
Florschütz, P.A. et al., 1120, 2741, 2755 (1-3)
Focke, H.C., 371 (1-3)

Heyde, N.M. & J.C. Lindeman, 25, 54, (1-4b), 123 (1-3)
Lanjouw, J., 1168, 1221 (1-3)
Lanjouw, J. & J.C. Lindeman, 323 (1-3), 349, 2416 (1-4b)
LBB, 14612 (1-4b)
Lindeman, J.C., 5785 (1-4b)
Lindeman, A.R.A. Görts-van Rijn et al., 26, 58 (1-4b); 98, 355 (1-8b)
Maguire, B., *24857* (1-2); 24858 (1-4b)
Mennega, A.M.W., 371 (1-3)
Mori, S.A. et al., 8522 (1-4b)
Splitgerber, F.L., 1136 (1-3)
Versteeg, G.M., 121, 153 (1-3)
Wullschaegel, H.R., cf. 1861 (1-3)

FRENCH GUIANA

Aublet, J.B.C.F. d', *s.n.* (1-4a); *s.n.* (1-8b)
Béna, P., 4204 (1-3)
Broadway, W.E., 383, 539, 662 (1-8b)
Granville, J.J. de, B-69 (1-8b)
Hallé, F., 847 (1-8b)
Leblond, J.B., 35 (1-4b); 36 (1-8b)
Lemée, A.M.V., s.n. (1-8b)
Leprieur, F.R., 237 (1-8b); 238 (1-4a); 1833, 1834, 1838, 1839, 1840 (1-8b)
Mélinon, M., 134, 136, 141, 230 (1-4b), 1865 (1-8b), s.n. (1-4b)
Perrottet, G.S., *259*, *260*, 261 (1-8b); 262 (1-4b); s.n (1-8b)
Poiteau, P.A., 1826 (1-8b); s.n. (1-4b)
Prévost, M.F., 804, 1165 (1-4a)
Skog, L. & C. Feuillet, 7445 (1-4a)
Soubirou, G., s.n. (1-4a)
Wachenheim, H., 479 (1-4b)

Krameriaceae

GUYANA

Davis, T.A.W., 873 (1-1); 1701 (1-2)
Harrison, S.G., 1320 (1-1)
Hoffman, B., 1063 (1-1)
Jansen-Jacobs, M.J. et al., 473 (1-2); 578 (1-1); 4596 (1-1); 5041 (1-2)
Irwin, H.S., 581, 885 (1-1)
Maas, P.J.M. et al., 3655 (1-1)
Schomburgk, Rob., I 201S, II 496 (1-2)
Schomburgk, Rich., 570 (1-2)
Wilson-Browne, G., 30 (1-2)

INDEX TO SYNONYMS, NAMES IN NOTES AND SOME TYPES

Vochysiaceae

Salvertia thyrsiflora Pohl = 4-1
Schuechia Endl. = 2
  brasiliensis Endl. ex Walp. = 2-4, type 2
  ecalcarata (Mart.) Warm. = 2-4
Strukeria Vell. = 5
  oppugnata Vell., type 5
Vochisia Juss. = 5
  guianensis (Aubl.) Lam. = 5-7
  tetraphylla (G.Mey.) Stone & Freeman = 5-14
Vochya Vell. ex Vandelli = 5
Vochysia arcuata Garcke = 5-14
  curvata Klotzsch = 5-3
  elongata Pohl = 5-16
  ferruginea (Mart.) Standl. = 5-5
  lucida Klotzsch = 5-6
  melinonii Beckmann = 5-7
  opaca Pohl ex Warm. = 5-16
  paraensis Huber ex Ducke = 5-7
  surinamensis Stafleu var. inflata Stafleu, see note 5-13
  tucanorum var. *elongata* (Pohl) Warm. = 5-16
  tucanorum var. fastigiata, see note 5-16
    var. *microphylla* Warm. = 5-16

Euphroniaceae

Euphronia hirtelloides Mart., type 1
Lightia R.H.Schomb. = 1
  guianensis R.H.Schomb. = 1-1, type 1

Trigoniaceae

Hoeffnagelia Neck. = 1
Mainea Vell. = 1
  racemosa Vell., type 1
Nuttallia Spreng. = 1
  villosa Spreng., type 1
Trigonia cepo Cambess. = 1-8b
  hypoleuca Griseb. var. *pubescens* Warm. = 1-3
  kaieteurensis Maguire = 1-4a
  macrocarpa Benth. = 1-8a
  macrostachya Klotzsch = 1-8b
  microcarpa Sagot ex Warm. = 1-4b

mollis DC. = 1-8b
parviflora Benth. = 1-4b
spruceana Benth. ex Warming, see doubtful species
villosa Aubl. var. *angustifolia* Benth. = 1-8b
var. *cuneata* DC. = 1-8b
var. *oblonga* DC. = 1-8b
var. *obtusata* DC. = 1-8b
xanthopila Garcke = 1-3

Krameriaceae

Krameria ixina L. = 1-1

INDEX TO VERNACULAR AND TRADE NAMES

(T.n. means mentioned as trade name in the WOOD AND TIMBER chapter)

Vochysiaceae

Alphabetic list of families of series A occurring in the Guianas

Defined as in Cronquist, 1981, and numbered in his sequence, with alternative names. Those published, with chronological fascicle number and year.

Family	No.		Fascicle
Abolbodaceae			
(see Xyridaceae	182)		15. 1994
Acanthaceae	156		
(incl. Thunbergiaceae)			
(excl. Mendonciaceae	159)		
Achatocarpaceae	028		
Agavaceae	202		
Aizoaceae	030		
(excl. Molluginaceae	036)		
Alismataceae	168		
Amaranthaceae	033		
Amaryllidaceae			
(see Liliaceae	199)		
Anacardiaceae	129		19. 1997
Anisophylleaceae	082		
Annonaceae	002		
Apiaceae	137		
Apocynaceae	140		
Aquifoliaceae	111		
Araceae	178		
Araliaceae	136		
Arecaceae	175		
Aristolochiaceae	010		20. 1998
Asclepiadaceae	141		
Asteraceae	166		
Avicenniaceae			
(see Verbenaceae	148)		4. 1988
Balanophoraceae	107		14. 1993
Basellaceae	035		
Bataceae	070		
Begoniaceae	065		
Berberidaceae	016		
Bignoniaceae	158		
Bixaceae	059		
(incl. Cochlospermaceae)			
Bombacaceae	051		
Bonnetiaceae			
(see Theaceae	043)		
Boraginaceae	147		
Brassicaceae	068		
Bromeliaceae	189	p.p.	3. 1987
Burmanniaceae	206		6. 1989
Burseraceae	128		
Butomaceae			
(see Limnocharitaceae	167)		
Byttneriaceae			
(see Sterculiaceae	050)		
Cabombaceae	013		
Cactaceae	031		18. 1997
Caesalpiniaceae	088	p.p.	7. 1989
Callitrichaceae	150		
Campanulaceae	162		
(incl. Lobeliaceae)			
Cannaceae	195		1. 1985
Canellaceae	004		
Capparaceae	067		
Caprifoliaceae	164		
Caricaceae	063		
Caryocaraceae	042		
Caryophyllaceae	037		
Casuarinaceae	026		11. 1992
Cecropiaceae	022		11. 1992
Celastraceae	109		
Ceratophyllaceae	014		
Chenopodiaceae	032		
Chloranthaceae	008		
Chrysobalanaceae	085		2. 1986
Clethraceae	072		
Clusiaceae	047		
(incl. Hypericaceae)			
Cochlospermaceae			
(see Bixaceae	059)		
Combretaceae	100		
Commelinaceae	180		
Compositae			
(= Asteraceae	166)		
Connaraceae	081		
Convolvulaceae	143		
(excl. Cuscutaceae	144)		
Costaceae	194		1. 1985
Crassulaceae	083		
Cruciferae			
(= Brassicaceae	068)		
Cucurbitaceae	064		
Cunoniaceae	081a		
Cuscutaceae	144		
Cycadaceae	208		9. 1991
Cyclanthaceae	176		
Cyperaceae	186		
Cyrillaceae	071		
Dichapetalaceae	113		
Dilleniaceae	040		
Dioscoreaceae	205		
Dipterocarpaceae	041a		17. 1995
Droseraceae	055		
Ebenaceae	075		
Elaeocarpaceae	048		
Elatinaceae	046		
Eremolepidaceae	105a		
Ericaceae	073		
Eriocaulaceae	184		
Erythroxylaceae	118		

Euphorbiaceae	115	
Euphroniaceae	123a	21. 1998
Fabaceae	089	
Flacourtiaceae	056	
(excl. Lacistemaceae	057)	
(excl. Peridiscaceae	058)	
Gentianaceae	139	
Gesneriaceae	155	
Gnetaceae	209	9. 1991
Gramineae		
(= Poaceae	187)	8. 1990
Gunneraceae	093	
Guttiferae		
(= Clusiaceae	047)	
Haemodoraceae	198	15. 1994
Haloragaceae	092	
Heliconiaceae	191	1. 1985
Henriquesiaceae		
(see Rubiaceae	163)	
Hernandiaceae	007	
Hippocrateaceae	110	16. 1994
Humiriaceae	119	
Hydrocharitaceae	169	
Hydrophyllaceae	146	
Icacinaceae	112	16. 1994
Hypericaceae		
(see Clusiaceae	047)	
Iridaceae	200	
Ixonanthaceae	120	
Juglandaceae	024	
Juncaginaceae	170	
Krameriaceae	126	21. 1998
Labiatae		
(= Lamiaceae	149)	
Lacistemaceae	057	
Lamiaceae	149	
Lauraceae	006	
Lecythidaceae	053	12. 1993
Leguminosae		
(= Mimosaceae	087)	
+ Caesalpiniaceae	088)	p.p. 7. 1989
+ Fabaceae	089)	
Lemnaceae	179	
Lentibulariaceae	160	
Lepidobotryaceae	134a	
Liliaceae	199	
(incl. Amaryllidaceae)		
(excl. Agavaceae	202)	
(excl. Smilacaceae	204)	
Limnocharitaceae	167	
(incl. Butomaceae)		
Linaceae	121	
Lissocarpaceae	077	
Loasaceae	066	
Lobeliaceae		
(see Campanulaceae	162)	
Loganiaceae	138	
Loranthaceae	105	
(excl. Viscaceae	106)	
Lythraceae	094	
Malpighiaceae	122	
Malvaceae	052	
Marantaceae	196	
Marcgraviaceae	044	
Martyniaceae		
(see Pedaliaceae	157)	
Mayacaceae	183	
Melastomataceae	099	13. 1993
Meliaceae	131	
Mendonciaceae	159	
Menispermaceae	017	
Menyanthaceae	145	
Mimosaceae	087	
Molluginaceae	036	
Monimiaceae	005	
Moraceae	021	11. 1992
Moringaceae	069	
Musaceae	192	1. 1985
(excl. Strelitziaceae	190)	
(excl. Heliconiaceae	191)	
Myoporaceae	154	
Myricaceae	025	
Myristicaceae	003	
Myrsinaceae	080	
Myrtaceae	096	
Najadaceae	173	
Nelumbonaceae	011	
Nyctaginaceae	029	
Nymphaeaceae	012	
(excl. Nelumbonaceae	010)	
(excl. Cabombaceae	013)	
Ochnaceae	041	
Olacaceae	102	14. 1993
Oleaceae	152	
Onagraceae	098	10. 1991
Opiliaceae	103	14. 1993
Orchidaceae	207	
Oxalidaceae	134	
Palmae		
(= Arecaceae	175)	
Pandanaceae	177	
Papaveraceae	019	
Papilionaceae		
(= Fabaceae	089)	
Passifloraceae	062	
Pedaliaceae	157	
(incl. Martyniaceae)		
Peridiscaceae	058	
Phytolaccaceae	027	
Pinaceae	210	9. 1991